Ocean Zoning

Ocean Zoning

Making Marine Management More Effective

Tundi Agardy

publishing for a sustainable future
London • Washington, DC

First published in 2010 by Earthscan

Copyright © Dr Tundi Spring Agardy, 2010

The moral right of the author has been asserted.

All rights reserved. No part of this publication may be reproduced, stored in a retrieval system, or transmitted, in any form or by any means, electronic, mechanical, photocopying, recording or otherwise, except as expressly permitted by law, without the prior, written permission of the publisher.

Disclaimer
The designations employed and the presentation of the material in this publication do not imply the expression of any opinion whatsoever on the part of the United Nations Environment Programme concerning the legal status of any country, territory, city or area or of its authorities, or concerning delimitation of its frontiers or boundaries. Moreover, the views expressed do not necessarily represent the decision or the stated policy of the United Nations Environment Programme, nor does citing of trade names or commercial processes constitute endorsement.

Earthscan Ltd, Dunstan House, 14a St Cross Street, London EC1N 8XA, UK
Earthscan LLC, 1616 P Street, NW, Washington, DC 20036, USA

Earthscan publishes in association with the International Institute for Environment and Development

For more information on Earthscan publications, see www.earthscan.co.uk
or write to earthinfo@earthscan.co.uk

ISBN: 978-1-84407-822-6

Typeset by JS Typesetting Ltd, Porthcawl, Mid Glamorgan
Cover design by Clifford Hayes

A catalogue record for this book is available from the British Library

Library of Congress Cataloging-in-Publication Data

Agardy, Tundi.
  Ocean zoning : making marine management more effective / Tundi Agardy.
    p. cm.
  Includes bibliographical references and index.
  ISBN 978-1-84407-822-6 (hardback)
 1. Ocean zoning. 2. Marine ecology. 3. Marine biology. 4. Marine pollution.
I. Title.
  GC1018.5.A43 2010
  333.95'616--dc22
                              2010002005

At Earthscan we strive to minimize our environmental impacts and carbon footprint through reducing waste, recycling and offsetting our CO_2 emissions, including those created through publication of this book. For more details of our environmental policy, see www.earthscan.co.uk.

This book was printed in the UK by MPG Books, an ISO 14001 accredited company. The paper used is FSC certified.

Contents

List of Photographs, Figures and Colour Plates		*vii*
Preface		*ix*
1	Introduction	1
2	Marine Management Challenges: How Ocean Zoning Can Help Overcome Them	23
3	Ocean Zoning Steps	41
4	Zoning within the Great Barrier Reef Marine Park (Australia)	59
5	Various Incarnations of Ocean Zoning in New Zealand	75
6	Zoning Efforts in United Kingdom Waters	83
7	Zoning Undertaken by the OSPAR Countries of the Northeast Atlantic	97
8	Zoning the Asinara Marine Park of Italy	111
9	Possibilities for Holistic Zoning of the Mediterranean Sea	121
10	Integrated Coastal Management, MPA Networks and Large Marine Ecosystems in Africa	133
11	Zoning within Marine Management Initiatives in British Columbia, Canada	147
12	Marine Spatial Planning in the US	155
13	Principles Underscoring Ocean Zoning Success	165
14	Implementing Ocean Zoning	175
15	Conclusions	191
Annex: IUCN Protected Area Categories		199
Recommended Further Reading		209
Index		213

List of Photographs, Figures and Colour Plates

Chapter photographs

All images by Tundi Agardy, unless indicated otherwise.

Preface	The author research-diving in Palmyra Atoll	xi
Chapter 1	Signpost at Land's End (Cornwall, UK) and the sea's beginning	1
Chapter 2	Pelican in the port of Walvis Bay, Namibia	23
Chapter 3	Mangrove conservation site in Tikina Wai, Fiji	41
Chapter 4	Reef flat in the Great Barrier Reef Marine Park	59
Chapter 5	Is the beach considered Crown land?	75
Chapter 6	The Isles of Scilly, off southwestern England	83
Chapter 7	Offshore wind farm in Danish North Sea waters	97
Chapter 8	Placid and pristine Asinaran waters	111
Chapter 9	Dropoff near Tavolara, Sardinia – an example of the many highly productive and diverse areas of the Mediterranean Sea	121
Chapter 10	Fisherman poling through Mafia Island's shallows (Tanzania)	133
Chapter 11	Vancouver Island seascape, near Campbell Island, British Columbia	147
Chapter 12	Marine conservationists meeting to plan protection for Palmyra Atoll, a US Territory in the Line Islands	155
Chapter 13	Some would call this Fijian Island paradise	165
Chapter 14	Patmos Island landscape, Greece	175
Chapter 15	Dhow sailing in the Indian Ocean off Zanzibar, East Africa	191
Recommended Further Reading	The author in the Galapagos Islands Marine Park	209

Figures

1.1	Schematic of coastal system	3
1.2	Algerian beach scene exemplifying some potentially conflicting uses of the marine and coastal environment	5

4.1	San Andrés Zoning Plan within Seaflower Marine Protected Area	70
6.1	Limits of United Kingdom internal waters, Fishery Conservation Zone and continental shelf	85
6.2	Division of various management authorities under the Marine and Coastal Access Bill	86
7.1	Map of the OSPAR area	100
7.2	Marine biological valuation map of the Belgian Exclusive Economic Zone	103
8.1	Asinara Marine National Park, located off the northwest coast of Sardinia (Italy)	113
8.2	Structuring of information used in the Asinara MPA zoning process	117
8.3	Concordance maps for the protection scenarios in the Asinara MPA	118
9.1	Countries with watersheds draining into the Mediterranean Sea	123
9.2	The limits of the Pelagos Sanctuary for the protection of marine mammals in the Ligurian Sea	130
10.1	West African countries in the RAMPAO region	134
10.2	The Bijagos Biosphere Reserve in Guinea Bissau	135
10.3	Location of Namibian coastal parks and the Namibian Islands MPA	144
11.1	Map of British Columbia's waters, showing ecosections	149
11.2	Limits of the PNCIMA region	150
14.1	Relationship between marine zoning and other spatial management initiatives	188

Colour plates

- 4.1 The Great Barrier Reef Zoning Plan
- 6.1 Irish Sea Pilot irreplaceability of planning units
- 7.1 Example of a structural zoning map produced by GAUFRE, with zones superimposed on bathymetry
- 7.2 Ecosystem-based management plan for the Barents Sea
- 8.1 Biocenotic (habitat) map of Asinara Island's waters
- 9.1 Distribution of marine protected areas in the Mediterranean Basin
- 10.1 Mafia Island Marine Park boundaries and zonation
- 12.1 Nine Regional Planning Areas proposed for the US and corresponding minimum state representation

Preface

Chuck Birkeland

The author research-diving in Palmyra Atoll

I've long felt that sense of place matters in marine conservation. The wide blue ocean realm is far too vast for people to be able to connect with; most marine animals are too wide-ranging to give people cause to consider them 'theirs'. Without such associations, the global ocean reverts to being out of sight, out of mind, and ocean degradation remains a low priority for the public and for decision-makers.

In planning marine protected areas, conservationists are able to root marine issues to a specific place, and in so doing, engage the public and decision-makers in a concrete set of measures aimed to protect that place. But when we deal with the wider ocean, the meandering coasts and the wetlands that link land to sea, we seem to be lost in the enormity of it all.

We need a paradigm shift, or at the very least a change in tack. Comprehensive ocean zoning allows us to establish a sense of place within a logical strategic framework for considering and accommodating ocean uses.

At its simplest, ocean zoning puts our knowledge of ecosystems and our impact on it on the table for all to see. And what a powerful motivator that is indeed. For, to paraphrase an oft-invoked old saying, we cannot love what we do not know. Contrary to what is evoked by Richard Wilbur's poetic lines ...

> *All that we do is touched with ocean*
> *Yet we remain on the shore of what we know*

... we do in fact know quite a lot about the ocean and need to communicate what we know in a way that the public, the decision-makers and all ocean stakeholders can understand and accept.

A zoning approach that focuses on ecosystem function and the ecosystem services that are so crucial for life on Earth (including our own) could prevent the many mistakes of the past that resulted either from shying away from managing ocean uses (too big, too complicated, commons dilemmas, etc.) or from a fixation on structure over function.

A medical analogy may help explain what I mean. We have been too long obsessed with the way marine ecosystems are structured and less concerned with how their functioning is being affected by human activities both on land and at sea. It reminds me of the state of medicine centuries ago, when physicians focused solely on anatomy, and physiology hadn't yet matured to be of practical use.

If our interventions are going to help the dying ocean patient, we will have to do two things, and do them well: 1) practise triage by focusing first and foremost on what is most ecologically critical – the 'vital organs' of the marine system – and 2) address root causes of damage to these vital organs, instead of applying salves to alleviate symptoms.

To accomplish this shift from a preoccupation with ocean structure to a more physiological approach to ocean management, the most powerful management tool available may well be comprehensive ocean zoning. Zoning allows us to locate and protect critical areas from the real risks they face, and allows us to accommodate appropriate uses in all ocean places.

Place matters to planners, who align regulations with space to show on maps what is allowed where; to managers, who direct their operations site-by-site; to enforcement agents, who ground legitimate use to boundaries, thereby determining what is illegitimate or illegal; and to conservationists, who flag certain habitats as critically important for ecosystems and for species.

Place also matters to the rest of us – there are sacred spaces in all cultures and landscapes that soothe the mind and spirit. Comprehensive ocean zoning allows us to be cognizant of the immense value attached to coastal and marine places and to protect them as effectively and strategically as possible.

It is thrilling to see the sudden emergence of interest in marine spatial planning and ocean zoning. The tales this book provides of this emerging field are not meant to be comprehensive – they merely illustrate what in my own opinion are key concepts. There are many other examples of ocean zoning that have not been described herein, and each deserves our attention. And despite the efforts of the pioneering few who have tried its application, there remains trepidation in using ocean zoning to its full potential. It is clearly never easy to buck the status quo – even if ocean zoning may well drive the very paradigm shift we've all been awaiting. I thus have nothing but the utmost respect for all those who have catalysed ocean zoning efforts around the world, since applying an unproven methodology to meet the ever-growing challenges of protecting oceans and the valuable services they provide requires a modicum of bravery.

And speaking of utmost respect, I would like to acknowledge the many unseen forces at play in conceiving and producing this book. In a field frenetic with progress and new developments each and every day, it has been a challenge to stay current. Many of my friends and colleagues in marine conservation thoughtfully fed me news of marine spatial planning and zoning efforts around the world, including my co-conspirators in Mediterranean marine zoning Giuseppe Notarbartolo di Sciara and Ferdinando Villa; MPA News Editor-in-Chief and my co-editor at MEAM John Davis; Andy Hooten, with whom I've had the good fortune to work in many distant places; Ole Vestergaard, Jackie Alder and Richard Kenchington, who provided both UNEP support and their own insights; coastal planning genius Jim Dobbin; ocean zoning pioneer John Ogden; Australian zoning guru Jon Day; and my brilliant, iconoclastic good friend Paul Dayton. Forest Trends Director Michael Jenkins, its MARES Program, and the entire Forest Trends family supplied much support and encouragement. Earthscan has been a joy to work with, and editor Tim Hardwick is stellar.

Last but not least, my family deserves many thanks – notably my sister Adrienne Miller, who diligently reads everything I write though it holds absolutely no interest for her, and my wonderful husband Josh and children Alex, Sophie, and Christopher, as they all had to suffer through far too many bouts of irritability and distractedness on my part during the course of the year it took to write this tome.

1
Introduction

Signpost at Land's End (Cornwall, UK) and the sea's beginning

The state of the ocean environment

The global ocean is ailing.

With each passing day comes news of further degradation, continued overexploitation, heightened conflict, the ravages of climate change and even unanticipated environmental issues concerning our ocean and coasts. And while there are small-scale success stories out there, the scramble is on for a new paradigm and new way of doing business – the ocean management business.

The continuing decline in the health of the global ocean and its coasts comes with risk to human well-being everywhere. While coastal and marine ecosystems are dynamic, in many cases they are now undergoing more rapid change than at any time in their history (Millennium Ecosystems Assessment (MEA), 2005). Transformations have been physical, as in the dredging of waterways, infilling of wetlands and construction of ports, resorts and housing developments, and they have been biological, as has occurred with declines in abundances of marine organisms such as sea turtles, marine mammals, seabirds, fish and marine invertebrates (Halpern et al., 2008; Worm et al., 2006). The dynamics of sediment transport and erosion-deposition have been altered by land and freshwater use in watersheds; the resulting changes in hydrology have greatly altered coastal dynamics. These impacts, together with chronic degradation resulting from land-based and marine pollution, have caused significant ecological changes and an overall decline in many ecosystem services (MEA, 2005).

Human pressures on coastal resources compromise the delivery of many ecosystem services crucial to the well-being of coastal peoples and national economies. Stocks of coastal fishery species, like most offshore fisheries, have been severely depleted. In the latest estimate by the Food and Agricultural Organization of the United Nations, over 80 per cent of the world's commercially fished stocks were found to be at capacity or overexploited (FAO, 2009). A recent review of fisheries management across the globe finds that it is difficult to identify even a single coastal nation that is not under the influence of overcapacity of fishing fleets or perverse subsidies for fisheries development. And while there is much disagreement about whether fisheries management can keep abreast of the increasing pressures to supply fisheries products for consumption, to support agriculture and even to provide fertilizers for landscaping, even conservative fisheries managers largely agree that better management is needed (Worm et al., 2009).

Depletions in fisheries stocks not only cause scarcity in resource availability (as well as substantial wealth inequities in many parts of the world), but also change the productivity of coastal and marine food webs, impacting the delivery of other services important to mankind (Dayton et al., 1995; 2002). Such services include providing coastal developments protection from erosion and storm damage, and increasing the value of recreational and tourist experiences, among others.

Biological transformations are also coupled to physical transformations of the coastal zone. Habitat alteration continues to be pervasive in the coastal

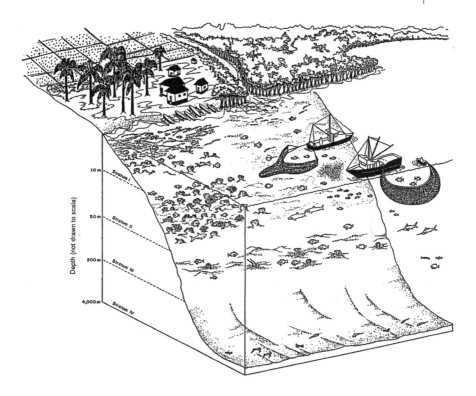

Figure 1.1 *Schematic of coastal system*

Source: Daniel Pauly

zone, and degradation of habitats both inside and outside these systems contributes to impaired ecological functioning. Similarly, human activities far inland, such as agriculture and forestry, impact coastal ecosystems when freshwater is diverted from estuaries, and when land-based pollutants enter coastal waters (nearly 80 per cent of the pollutant load reaching the oceans comes from terrestrial sources). These chemical transformations impact the viability of coastal systems and their ability to deliver services. Thus, changes to ecosystems and services occur as a function of land use, freshwater use and activities at sea, even though these land-freshwater-marine linkages are often overlooked (Figure 1.1).

Larger forces are also at play. Coastal areas are physically vulnerable: many areas are now experiencing increasing flooding, accelerated erosion and seawater intrusion into freshwater, and these changes are expected to be exacerbated by climate change in the future (IPCC, 2003). Such vulnerabilities are currently acute in low-lying mid-latitude areas, but both low-latitude areas and polar coastlines are becoming increasingly vulnerable to climate change impacts. Coral reefs and atolls, salt marshes and mangrove forests, and seagrasses will likely continue to be impacted by future sea-level rise, warming oceans and changes in storm frequency and intensity (Sale, 2008). Even open ocean systems are under severe threat from climate change impacts, including the ocean acidification and oxygen depletion that occurs as ocean waters warm (Brierley and Kingsford, 2009). For instance, some models predict that oxygen levels in the seas will decline 6 per cent for every 1 degree increase in temperature, leading to accelerated growth in 'dead zones' across the globe. These areas of low or no oxygen cannot continue to support marine life, and the expansion of these dead zones will likely have a profound effect not only on marine fisheries but on human health as well.

At the same time, the incidence of disease and emergence of new pathogens is on the rise, and in many cases will have significant human health consequences (NRC, 1999; Rose et al., 2001). Episodes of harmful algal blooms are increasing in frequency and intensity, affecting both the resource base and humans living in coastal areas more directly (Burke et al., 2001; Epstein and Jenkinson, 1993). Effective measures to address declines in coastal health and productivity remain few and far between, and are often too little, too late. Restoration of coastal habitats, although practised, is generally so expensive that is remains a possibility only on the small scale or in highly developed countries. Education about these issues is lacking, and therefore this crisis may rest in large part on our inability to communicate what is happening and why.

Dependence on coastal zones is increasing around the world, even as costs of rehabilitation and restoration of degraded coastal ecosystems are on the rise. In part this is because population growth overall is coupled with increased degradation of terrestrial areas (fallow agricultural lands, reduced availability of freshwater, desertification and armed conflict contributing to decreased suitability of inland areas for human use). Resident populations of humans in coastal areas are growing, but so are in-migrant (internal migrant) and tourist populations (Burke et al., 2001), while wealth inequities rise in part from the tourism industry and private landowners decreasing access to coastal regions and resources for a growing number of humans (Creel, 2003; McCay, 2008). Nonetheless, local communities and industries continue to exploit coastal resources of all kinds, including fisheries resources; timber, fuelwood and construction material; oil, natural gas, strategic minerals, sand and other non-living natural resources; and genetic resources. In addition, people increasingly use ocean space for shipping, security zones, renewable energy, recreation, aquaculture and habitation. The photo in Figure 1.2 shows a small stretch of coastline in Algeria, just one example of these disparate and potentially conflicting uses of coastal and marine ecosystems; these and other uses are the focus of zoning initiatives.

Coastal zones provide far-reaching and diverse job opportunities, and in many countries represent the geographic areas having the greatest contribution to GDP or GNP. Income generation and human well-being, as measured by various economic and social parameters, are currently higher on the coasts than inland across the globe.

Despite their value to humans, coastal and marine systems and the services they provide are becoming increasingly vulnerable. Although the thin strip of

Figure 1.2 *Algerian beach scene exemplifying some potentially conflicting uses of the marine and coastal environment*

Source: T. Agardy

coastal land at the continental margins and within islands accounts for only 5 per cent of the earth's land area, nearly 40 per cent of the global population lives in the coastal zone (CIESIN, 2003, using 2000 census data). Population density in coastal areas is close to 100 people per km^2 compared to inland densities of 38 people per km^2 in 2000 (MEA, 2005). Though many earlier estimates of coastal populations have presented higher figures (in some cases, nearly 70 per cent of the world population was cited as living within the coastal zone), previous estimates used much more generous geographic definitions of the coastal area and may be misleading (Tibbetts, 2002).

In general, management of coastal resources and human impacts on these areas is insufficient or ineffective, leading to conflict, decreases in services and decreased resilience of natural systems to changing environmental conditions. Inadequate fisheries management persists, often because decision-makers are unaware when marine resource management is ineffective, while coastal zone management rarely addresses problems of land-based sources of pollution and degradation (Kay and Alder, 2005). Funds are rarely available to support management interventions over the long term, resources become overexploited and then unavailable, and conflicts increase.

A new paradigm, or at the very least, a substantial ramping up of truly effective management, is badly needed. It is unlikely that old management tools and approaches will be sufficient to meet these ever-increasing, and sometimes newly emerging, challenges. But what is the top priority for a better ocean future?

Like many other assessments and publications before and since, the 2006 National Academy of Sciences book entitled *Increasing Capacity Building for Stewardship of Oceans and Coasts: A Priority for the 21st Century* (NRC, 2006) identified fragmentation of management as one of the most pervasive and prominent obstacles to effective marine management. Ocean zoning by its very nature overcomes fragmentation – obligating the managers of all the various sectors using marine resources and ocean space to think strategically and plan for sustainable use.

In facing a climate-changed and challenged future, which will have over 8 billion human inhabitants to house and feed, getting the management right is imperative. Only healthy and well-functioning ecosystems will be able to adapt to a changing world and continue to provide the goods and services that maintain life on earth. And only regulations and rules that are derived through the active participation of those who will be affected will be accepted, with a minimum of conflict and risk to national security. The strategic planning and adaptability inherent in zoning may be the single most important weapon in our armoury to achieve effective, efficient and equitable management of resources – a battle we have been waging, and losing, for quite some time now.

Ocean zoning to meet ocean management challenges

Zoning is a set of regulatory measures used to implement marine spatial plans – akin to land-use plans – that specify allowable uses in all areas of the target ecosystem(s). Different zones accommodate different uses, or different levels of use. As in municipal zoning, regulations address prohibitions or permitted uses, or both. All zoning plans are portrayed on maps, since the regulations are always area-based.

Ocean zoning has been repeatedly brought up as having much potential, as managers have struggled to slow or halt coastal degradation and overexploitation of marine resources. More often, government agencies, conservationists and planners have skirted ocean zoning without actually invoking the term. Instead, they speak of comprehensive ocean planning, marine spatial management, place-based conservation and the like. Although the ocean zoning concept itself is just a natural extension of what we do on land, for some reason it is feared. Thus comprehensive ocean zoning efforts have largely remained in the realm of theory, in part because of this fear.

Marine spatial management really began with indigenous coastal and island cultures and their marine tenure and taboo practices, and then became more widespread (and legitimate, in the eyes of some) when formal marine protected areas started to be established. Then, when practitioners realized that few marine protected areas were meeting broad scale conservation objectives, and that an ad hoc, one-off approach would not lead to effective large-scale conservation (Allison et al., 1998), the concept of marine protected area networks emerged as a way to strategically plan marine protected areas with the hope that the whole would be greater than the sum of its parts (Roberts et al., 2003). A system or network that links these protected areas has a dual nature: connecting physical sites deemed ecologically critical (ecological networks), and linking people and institutions in order to make effective conservation possible (human networks). Networks or systems of marine protected areas have great advantages in that they spread the costs of habitat protection across a wider array of user groups and communities while providing benefits to all.

While networks are a step in the right direction, even strategically planned networks do not necessarily lead to effective marine conservation at the largest scale (Christie et al., 2002). Recognizing that more was needed than marine protected area networks, planners began to explore the concept of marine corridors and broader spatial management. Essentially, a corridor uses a marine protected area network as a starting point and determines through conservation policy analysis which threats to marine ecology and biodiversity cannot be addressed through a spatial management scheme. The connections between the various marine protected areas in a network are maintained by policy initiatives or management reform in areas outside the protected areas. In such corridors (or regional planning initiatives by any other name) marine policies are directed not at the fixed benthic and marine habitat that typically is the target for protected area conservation, but rather at the water quality in the water column and the condition of marine organisms within it. Corridor concepts provide a way for planners and decision-makers to think about the broader ocean context in which protected areas sit and to develop conservation interventions that complement spatial management techniques like marine protected areas and networks. Corridors and regional planning efforts are few and far between, however, and most marine conservation still occurs through a piecemeal, almost desperate, process without large-scale visioning and coordinated strategic action.

But in the minds of many, a real quantum leap in conservation effectiveness occurs when planners scale up from marine protected area (MPA) networks and corridor concepts to full-scale ocean zoning (Agardy, 2007; 2009). Ocean zoning provides many benefits over smaller scale interventions: it can help overcome the shortcomings of MPAs and MPA networks in moving us towards sustainability; it is based on a recognition of the relative ecological importance and environmental vulnerabilities of different areas; it allows harmonization with terrestrial land-use planning; it can help better articulate private sector roles and responsibilities and maximize private sector investment by allowing free market principles to work in concert with government protections; and it moves us away from the terrestrial focus of traditional integrated coastal management efforts to more effective, integrated and holistic environmental management that fully includes uses of, and impacts on, the oceans.

Ocean zoning is a planning tool that comes straight out of the land-use planning methodologies developed in the 1970s and used at the municipal, county, state, regional and national levels. As on land, it allows a strategic allocation of uses based on a determination of an area's suitability for those uses, and reduction of user conflicts by separating incompatible activities. There are generally two components to an ocean zoning plan: 1) a map that depicts the zones, and 2) a set of regulations or standards applicable to each type of zone created (Courtney and Wiggin, 2003). While the concept of wholesale zoning of the oceans has been slow to gain public acceptance, it is increasingly popular with marine resource managers and conservationists, and many countries are now experimenting with the approach at national, state or provincial, or local scales (Hendrick, 2005).

Many countries that struggle with how to accommodate multiple uses of ocean space and resources are now experimenting with larger scale zoning, usually referred to as 'marine spatial planning'. According to participants at a meeting hosted by the United Nations Education, Scientific and Cultural Organization (UNESCO) in 2006, China is acting on legislation it passed specifically for zoning in its territorial sea, and the United Kingdom has drafted national legislation that will authorize marine spatial management. Both Germany and Belgium have taken steps to extend land-use planning utilizing zoning to the marine environment, and a working group of the Oslo-Paris (OSPAR) Convention is drafting guidelines for marine spatial planning throughout the northeast Atlantic. New Zealand government officials working on its new Ocean Policy have explored ocean zoning within the country's Exclusive Economic Zone (EEZ), and Vietnam and Mexico are both setting out to do comprehensive ocean zoning with the passage of enabling legislation. Australia has been a trendsetter with its zoning work in the Great Barrier Reef (see Chapter 4); this has spurred other comprehensive zoning efforts in South Australia (Day et al., 2007) and elsewhere on the continent. Likewise, the Canadian government is toying with the idea of ocean zoning as it implements its policies and programmes under the Oceans Act.

In the United States, some individual states have taken the lead on exploring the concept of zoning, making it likely that zoning plans will first be implemented at the state level, within the 3-mile jurisdictions that the states control. For instance, the Commonwealth of Massachusetts is exploring spatial management options for state waters, and has established itself as a national leader in thinking about ecosystem-based management and the role of zoning in it. This is in part a response to two developments that were viewed as possible threats to Massachusetts' coastal waters: 1) the siting of an liquefied natural gas terminal outside Boston, and 2) the proposal to develop an offshore wind farm in Nantucket Sound. The state of California is also exploring zoning within state waters as part of the activities being undertaken by the California Marine Life Protection Act. Rhode Island has also jumped on the zoning-state-waters bandwagon and is actively engaging its cadre of professionals – who so expertly developed the state's coastal management policy – to develop comprehensive zoning plans to 3 nautical miles offshore.

Whether ocean zoning in these state-led initiatives will be housed under the state coastal planning and management agencies (Office of Coastal Zone Management or similar), or whether new ocean zoning entities will be established, will depend on the capacity of existing agencies and the resources available to the state. With the increasing attention and interest in ocean zoning, it is possible that the federal government will follow the lead of the states and assess what kind of zoning should occur in federal waters. This would bring the US in step with the many maritime nations that are developing marine policies that embrace ocean zoning. A National Center for Ecological Analysis and Synthesis (NCEAS) group at the University of California, Santa Barbara, has been working on the topic since late 2005.

On the international front, the Intergovernmental Oceanographic Commission of UNESCO has convened several international workshops on spatial management (UNESCO, 2009). A major conclusion of these meetings was that human uses of ocean space often conflict with one another, and marine spatial planning is a logical response to organize human activities in space and time to minimize conflict.

Adapting zoning from land to ocean environments

Zoning is a tool developed for use on land, primarily in cities and communities, in order to foster uses appropriate to particular places and minimize conflict between incompatible uses. On land, zoning is defined as the legislative division of a community area of which only certain designated uses of land are permitted.

In the United States, zoning is commonly employed by local authorities to promote the public welfare. There, and elsewhere in the world, zoning regulations are used by municipalities to restrict the development of property within their borders. Throughout western societies, zoning is accomplished by local legislative bodies, which develop comprehensive zoning schemes that separate residential use from industrial development and often delineate areas for agricultural use as well. But even in other social systems lacking property rights, zoning-like allocation of uses fixed to certain areas of land have been practised for centuries.

According to a review of zoning issues brought before the US Supreme Court, zoning was predicated on traditional principles of common-law nuisance, and from that early and loosely defined start grew throughout the twentieth century (Hall et al., 1992). But as zoning became more commonplace, restrictions and regulations started to be challenged in the courts. Debate centerd on questions of due process, equal protection and taking private property without just compensation.

Since the mid-twentieth century, innovative zoning schemes have addressed social issues in suburban and urban areas of countries like the United States. For example, zoning restrictions have been utilized to promote aesthetic values, to preserve historical landmarks and districts, to safeguard culturally important areas and to manage growth. These modern adaptations have increasingly been challenged, and in the United States a number of land-use regulation cases have reached the penultimate judicial level of the Supreme Court.

Clearly a major issue in land-use zoning, and one that does not pertain to zoning at sea, is the question of the extent to which government can ascertain ownership of private land or control uses of it, when done in the name of the common good. Within the United States, many legal battles have arisen from zoning decisions related to land use, include easements (forfeiting property rights for certain uses, such as providing traditional access through private land to public places) and eminent domain (seizure of private land by the government in order to fulfil the common good). An easement is a property interest that allows the holder of the easement to use property that he or she does not own or possess. Even though eminent domain or condemnation cannot occur without compensation to the private landowner, this government power to take private land for public use has been a particularly contentious side issue of zoning in recent years.

How does ocean zoning then compare to zoning practised on land and under the control of local authorities? According to Barale et al. (2009), one fundamental difference has to do with mindset: land is planned by urbanization laws, with the sea as a visual asset only, while, in contrast, marine specialists view the land as an enemy.

Ocean zoning differs from land-use zoning in other more obvious ways: the watery portion of the coastal zone and marine realm can only accommodate a more restricted suite of uses. With rare exceptions the sea (water column and seafloor) are commons property and thus condemnation is unlikely to be required, though restricting traditional uses is certainly a facet of zoning plans. Unlike on land, fencing and signage are of no help whatsoever in informing the public about which zone they are in or might wish to go to. In addition, ocean zoning is likely to be dynamic, with zones moving year to year or even seasonally, in some cases.

Comprehensive Ocean Zoning (COZ) is even a step beyond. COZ differs from land zoning in the scale by which it is planned and the way that it is implemented across the mosaic of private property, common property, and use-restricted landscapes and seascapes. For whereas terrestrial zoning is usually small scale, often within the remit of municipal planning authorities, COZ must recognize the wide linkages across marine and coastal ecosystems, and systematically address uses of and impacts on the marine environment at the scale of entire regions. That said, many conservationists and planners have argued that emphasizing the similarities between zoning on land and COZ, as opposed to highlighting the differences, will allow for ocean planning efforts that are familiar to the body politic and thus more acceptable.

The importance of linking ocean zoning to land-use planning

Comprehensive ocean zoning that solely targets the sea in isolation of what is happening on the coasts and inland is likely a folly. According to the European Commission, it is crucial that there is continuity of activities on land and sea, requiring complementarity between marine and land strategies and planning, as well as coherence in implementation. Regional approaches to planning zoning can overcome the issues of fragmentation and insufficiency that plague many marine management efforts.

There are important precedents for regional zoning approaches, suggesting that strategic, large-scale planning does hold promise for more effective marine conservation (Agardy, 2005). One is the relatively recent coupling of coastal zone management with catchment basin or watershed management, as has occurred under the European Water Framework Directive and projects undertaken under the LOICZ (Land-Ocean Interactions in the Coastal Zone) initiative. These fully integrated initiatives, with affecting and affected parties taking part in the planning process, have resulted in lower pollutant loads and improved conditions in some estuaries (MEA, 2005).

Regional approaches utilizing MPA networks and systems, which may well presage later ocean zoning, are also being discussed and developed in the Mediterranean under the Barcelona Convention (Convention for the Protection of the Marine Environment and the Coastal Region of the Mediterranean), in North America under the auspices of the North American Commission on Environmental Cooperation, and at the national scale in countries ranging from Australia to the United States. Smaller regions such as the Gulf of Maine, shared by Canada and the US, are also focal points for regional cooperation, as demonstrated by the multilateral work undertaken as part of the United Nations Environment Programme (UNEP) Global Programme of Action (see www.gpa.unep.org) and the work of the Gulf of Maine Council (see www.gulfofmaine.org).

Zoning as an integrator

In order to plan and institute zoning schemes, planners must recognize connections, including the connections between different elements in an ecosystem, between land and sea, between humans and nature, and between uses of ocean resources or ocean space and the ability of ecosystems to deliver important goods and services.

Given these connections, a spatial management strategy that involves zoning requires a scaling up: from single-species fisheries management to management encompassing whole species assemblages; from looking at isolated drivers of change to considering all environmental and human impacts; from design of individual protected areas to planning MPA networks; from conservation of a fragment of habitat to comprehensive spatial management.

Some management authorities in different parts of the world have taken on a hierarchical planning approach, which begins at the large scale and focuses in on ever-smaller scales; others work to reconcile the disconnects between management sectors at various scales as they arise.

Many large comprehensive ocean zoning efforts tend to be top-down policy efforts. Unlike other large-scale management measures, ocean zoning not only allows managers to consider the big picture but also provides a detailed roadmap for how to implement at small scales, so that local managers need not wonder how to interpret policy language or find resources to realize needed changes.

An important factor in the bridging of scales is the human/social dimension of governance. Countries with a history of centralized government and government-dominated governance exhibit a different way of coordinating management from cultures in which governance extends to other, non-governmental institutions. And finally, the scale of the ecosystem or ecoregion under consideration is significant, such that the larger the geographic and sectoral scales, the greater the need for a hierarchical or integrated strategy to maintain linkages between scales.

Thus a critical step in being able to practise effective management through ocean zoning is to recognize such disconnects, anticipate them in planning, and take necessary measures to promote awareness, understanding and cooperation.

The planning process for zoning therefore acts as an integrator: forcing the bringing together of information, user groups and management entities. Comprehensive ocean zoning can also facilitate integration across biomes or realms, along the continuum from river basins, through coastal waters, and out to sea.

The efficacy of watershed management also has direct bearing on the scope and scale of challenges marine managers face. Freshwater ecosystems that are degraded or poorly managed contribute significantly to degradation of marine ecosystems, and the reduced productivity and loss of services that are a challenge for marine managers everywhere. Freshwater systems deliver pollutants to coastal waters, but the effects extend well beyond. Loss of estuarine habitats, often the result of poor watershed management, denies marine ecosystems nursery areas needed by many marine species for continued production. And alarmingly, there is evidence to suggest that degraded freshwater ecosystems drive benthic/pelagic decoupling in the marine environment, such that the benthos is unable to act as a carbon sink. This in turn may drive accelerated warming and degradation of not just the marine environment, but the planet's environment as a whole.

Regional cooperation to address issues of water use and allocation, as well as threats to freshwater systems originating from pollution, overfishing, and changes in riparian landscapes, coupled to similarly integrated management of adjacent marine areas, holds great promise for effectively managing aquatic systems. There are good examples of watershed/waterbasin management frameworks and institutions in existence around the world, from the Mekong River Commission (Vietnam, Thailand, Laos, Cambodia) to the International Commission for the Protection of the Danube River (Austria, Bosnia-Herzegovina, Czech Republic, Germany, Hungary, Moldova, Romania, Serbia, Slovakia and Ukraine), to the drought-parched Murray/Darling Basin in Australia (involving the states of South Australia and New South Wales). Choice of scale in which to frame big picture watershed management is critical (see MEAM, 2008).

In developing its five principles for marine spatial planning, the European Commission (EC, 2008) recognizes that not only is integration across scales important, but that it also must be done in a transparent manner that makes the process of integration understandable to the public. These five principles are:

1 Cross-border cooperation
2 Incorporating monitoring and evaluation
3 Strong data and knowledge base
4 Coordination within Member States – simplifying decision processes
5 Developing Marine Spatial Planning in a transparent manner

Clearly, large-scale, top-down, command and control form of management has its limitations, without effective local involvement at much smaller scales. Thus a key concept is that regional cooperation is not solely the purview of national governments and high-level agencies. The broader the participation in cooperative management, the greater its potential for success.

Finally, as any student of governance will know, establishing frameworks for cooperation and management do not guarantee success. The litmus test for success in integrating across scales is whether such frameworks are leading to demonstrable positive outcomes on the ground. Instituting comprehensive ocean zoning is one clear way to move beyond the theoretical or conceptual big picture planning inherent in marine spatial planning to the operational steps of truly effective management.

Differences between marine spatial planning and ocean zoning

Marine spatial planning (MSP) is a generic term describing the process leading to place-based marine management. UNESCO defines MSP as 'the public process of analysing and allocating the spatial and temporal distribution of human activities in marine areas to achieve ecological, economic, and social objectives that are usually specified through a political process' (UNESCO, 2009). The European Union states that MSP is a process that consists of data collection, stakeholder consultation and the participatory development of a plan, the subsequent stages of implementation, enforcement, evaluation and revision (EC, 2008). MSP is thus increasingly seen as a central component to effective marine management, or perhaps a new name for what agencies have been struggling to do all along. For example, marine spatial planning has been defined by the UK's Department of Environment, Food and Rural Affairs as 'a strategic plan for regulating, managing, and protecting the marine environment that addresses the multiple, cumulative, and potentially conflicting uses of the sea'.

Comprehensive ocean zoning (COZ) is one tool used by marine spatial planners to integrate management of various activities. Though zoning is one of the central components of MSP, contrary to public perception, the two are not one and the same. Crowder et al. (2006) provide a detailed discussion of MSP and ocean zoning's role in it, as do Ehler and Douvere in their series of publications under UNESCO's Marine Spatial Planning initiative (UNESCO,

2009). However, as their comprehensive review on MSP and zoning (*Visions for a Sea Change*) suggests, many MSP initiatives do indeed utilize ocean zoning at small scales or COZ at even larger scales (Ehler and Douvere, 2007). A 2008 special issue of *Marine Policy*, guest edited by Ehler and Douvere, provides further proof that many regions are looking to COZ to bolster management efforts.

That marine spatial planning is suddenly in vogue and being invoked as a major new thrust of national marine policy in many coastal nations can only be seen as a positive development – a sign that countries are committed to being more strategic and integrative in their approach to ocean management. However, many planners and resource managers feel that MSP is merely a new label for something that agencies have been trying to practise for decades. Twenty years ago this type of planning would have been called Integrated Coastal Management (ICM) or Integrated Coastal Zone Management (ICZM); decades prior it would have been called Regional Planning. As coastal planning pioneers like James Dobbin of Canada have pointed out, MSP has been practised for the last 40 years, and the principles of MSP are no different than those of 'bioregionalism', which emerged in western countries in the 1970s.

But another way to look at it is to consider marine spatial planning as the framework, or the enabling environment, that makes comprehensive zoning possible. It is the zoning that is new and different – an innovative new tool that could spell the difference between the development of endless plans (many of which become shelved) and an actual improvement of management on the ground. I would argue that MSP without subsequent (or parallel) ocean zoning of some type is incomplete management or, better put, is not taking advantage of the power of ocean zoning as a problem-solving tool. Although the scenario development that much MSP entails allows us to look to the future and evaluate various management options, at some point, 'planning' can divert energy and resources away from 'doing' – and zoning represents the doing to which MSP leads.

Ocean zoning is therefore a potentially powerful tool for integrating marine management at ecosystem, and larger, scales.

The contingent of marine planners and conservationists who see enormous potential in using zoning as a tool to integrate marine management and make it truly ecosystem-based is growing. Some feel that ocean zoning has a key role in fisheries management, saying that zoning provides a comprehensive framework to realign fishing industry incentives and at the same time maintains a healthier ecosystem (Sanchirico, 2009). Such analysts contend that zoning can overcome both sets of problems by resolving conflicts among and between interest groups and breaking the logjam that prevents ecosystem values from being recognized. Specifically, they view zoning as an instrument to strengthen politically weak groups and provide ownership-related incentives to all groups (Sanchirico et al., 2010).

Other nations and authorities see great value in zoning, even if few have instituted full-scale COZ. The state of Massachusetts in the US recently passed the Massachusetts Ocean Act, which implicitly calls for spatial management.

The Act states that: 'The secretary, in consultation with the ocean advisory commission ... and the ocean science advisory council ... shall develop an integrated ocean management plan, which may include maps, illustrations and other media. The plan shall: ... set forth the commonwealth's goals, siting priorities and standards for ensuring effective stewardship of its ocean waters held in trust for the benefit of the public ... [and] shall identify appropriate locations and performance standards for activities, uses and facilities allowed'. As expected, state planning under the Act (to 3 nautical miles out) involves zoning, providing dedicated space for ocean industries like aquaculture and wind energy, and having prohibitions for certain uses in specific zones.

Ocean zoning can thus provide many benefits over other tools of marine management: overcoming the shortcomings of small-scale protected areas; recognizing the relative ecological importance and environmental vulnerability of different areas; allowing harmonization with terrestrial land-use and coastal planning; better articulating private sector roles, responsibilities and market opportunities; minimizing conflict between incompatible uses; and moving us away from fragmented sectoral efforts towards integrated and effective ecosystem-based management (EBM) that fully includes all uses of, and impacts on, the oceans.

A vital step in the process of using COZ is establishing goals and objectives for zoning, as well as a strategic vision. This is best achieved through a participatory process (see Chapter 13 for more detail on participatory planning). Another crucial next step is information gathering and analysis. The types of data needed include habitat classification and distribution, habitat values and vulnerabilities, and existing and prospective uses of ocean space and resources. Using these data requires knowledge about the goals and objective, and is facilitated by Geographic Information Systems (GIS), Multivariate Spatial Analysis, and decision-support tools like MARXAN, MARXAN with Zones ('Zonae Cognito'), Vista and others.

There is some debate about whether legislation mandating zoning is a prerequisite for using zoning in marine spatial planning. This seems to vary from place to place, even within the confines of a single nation. In the United States, zoning was clearly called for in the run up to the passage of the Massachusetts Ocean Act, but in California, where zoning of state waters seems an inevitable eventuality, there was no stated mandate. On the other hand, COZ in Belgium has progressed despite the lack of a legal framework, suggesting that legislative mandates are not a prerequisite for being able to utilize the ocean zoning tool.

UNESCO's Ehler and Douvere have developed guidelines for MSP that elucidate methodologies for utilizing COZ. Their guidelines cover:

- Approaches to establish the **authority** that allows the development of marine spatial management in a participatory manner that integrates issues across sectors
- Setting up **planning and analysis** for marine spatial management that allows proactive, future-oriented management of oceans and coasts

- Types of **research, data and information** essential to conduct marine spatial management that addresses both important ecological and socioeconomic concerns
- Incentives, institutional arrangements and other considerations for successful **implementation** of marine spatial management
- Processes to conduct practical **stakeholder involvement** in the pre-planning, planning, implementation and evaluation phases of marine spatial management
- Methods for the **adaptation** of marine spatial management plans to changing circumstances, including climate change, new political priorities or economic conditions

Nonetheless, it is likely that the specific process for using COZ will differ from place to place. These refinements will reflect the legislative and regulatory framework already in place, the availability of data on ecosystems, services and uses, and the predilection of the decision-makers and public for understanding and accepting ocean zoning.

COZ has much potential, but remains largely untested at the scales in which it could be called truly 'comprehensive'. A paper by Ben Halpern and colleagues claims that ocean zoning can make management spatially more coherent, acting to address interactions between various activities that impact marine ecosystems, cumulative impacts over time, the delivery of ecosystem services and the trade-offs between various uses (Halpern et al., 2008).

Thus it seems that ocean zoning may be an inevitable future development as nations and communities struggle to better manage marine resources to safeguard their futures.

Why go beyond MSP to ocean zoning?

While understanding of oceans and their link to human well-being is improving with increased awareness and advances in technology, we face defeat after defeat in trying to implement marine management that actually works. And, though we can point to occasional successes in addressing threats to marine ecosystems, the fact that these successes are small scale, and are often short-lived, means that the search for a more robust method of managing our oceans must continue.

The oceans are simultaneously dying the silent death of a thousand cuts and at the same time are visibly and dramatically being impacted by catastrophic events. We are sophisticated enough in our knowledge to address both these problems: what is needed to halt or reverse chronic degradation, and what can be done to make oceans and coasts maximally resilient in the face of extreme events? Yet we seem incapable of doing it.

Comprehensive ocean zoning may well change that – allowing us to radically shift the way we manage oceans and coasts.

Across the globe, we know what areas are more important to protect than others, in order to keep ecosystems healthy and productive. We largely understand what threatens these important areas and how to counter those threats with the spate of well-developed management tools and policy instruments available to us. Yet government authorities have been horribly timid and about applying the knowledge and the tools, fearing public backlash and resistance from the various users of ocean resources and space.

We flounder at our own peril, since the condition of the oceans and coasts is inextricably tied to our own. While management authorities wait for additional scientific information, address problems in a fragmented and uncoordinated way and use outdated methods to counter crises as they flare up, the oceans continue their long and seemingly inevitable decline.

Ocean zoning can improve public understanding and reduce conflicts by displaying what we know about ecosystems, their value and their condition in graphic ways that people can get their heads around. Figuring out what uses are appropriate where, and what levels of use will keep ecosystems productive and healthy, provides us a much-needed change in tack.

Zoning is simple, straightforward, systematic and strategic. It involves the identification of ecologically vital areas (areas with a high concentration of ecosystem services are a good start), assessment of threats to the delivery of those services and development of zoning schemes that prohibit harmful uses while permitting other uses at levels ensuring sustainability. To some extent this is already being practised when closed areas are designated or management regimes for marine natural resources are created, such as the noteworthy 2009 declaration by the North Pacific Fisheries Council to close off all the Arctic areas to commercial fishing. But comprehensive ocean zoning would take this much further – allowing the coordination of piecemeal attempts to manage different resources, different areas and different users into a coherent and strategic whole. The resulting plans could be welcomed by users and conservationists alike, as they would clearly lay out what is known about marine ecosystem value, vulnerabilities and needed management. The greater transparency also creates opportunities for shared stewardship of ocean areas and resources.

Clarification of rights and creating conditions that allow for the sharing responsibilities for management will greatly improve our ability to reverse degradation and improve ocean health. And comprehensive ocean management in which the private and public sectors work hand in hand is infinitely more efficient and understandable to the public than the uncoordinated array of largely vapid measures we are currently using to manage our impacts on the seas.

That is not to say there are no downsides. Besides being a largely unproven methodology, there may be a kneejerk negative reaction to ocean zoning as open access is diminished and the somewhat mystical appeal of unlimited, unregulated ocean use is extinguished (see for instance Blaustein, 2005). Even highly dedicated marine conservationists may balk at the idea of parcelling the ocean into areas according to their human-use values. In the practical development of zoning plans, there will undoubtedly be difficulties as competing interests, who have attached their own values to certain areas, are not given time and means to find resolution.

Does ocean zoning really offer a radical new approach to ocean management?

To date, most marine conservation has not happened on the global or regional scale – it has occurred bit by bit, as a result of individuals, communities and institutions responding to a particular need at a particular site. Traditional marine conservation interventions have included marine protected areas, regulations to protect critical habitat of a species and fisheries restrictions pertaining to a threatened stock. The size of these responses is usually far too small to address the bigger (and growing) problems of unsustainable use of resources, indirect degradation of marine ecosystems and large-scale declines in environmental quality, such as those brought about by climate change, since virtually all the world's nearshore areas experience multiple threats that act simultaneously to degrade ecosystems. And because oceans are indeed the ultimate sink, and the fate of coastal waters is so strongly tied to the condition of coastal lands, rivers and estuaries, successful conservation means addressing not only marine use but land use as well, far up into watersheds (Kay and Alder, 2005).

The steps taken to develop and implement a zoning plan can be analogous to those taken in MPA planning. That is, once a region to be zoned is bounded (either by jurisdictional considerations or by ecosystem boundaries), the areas within can be examined according to their ecological significance, value and use, and condition. Zoning patterns that emerge from a planning process will be based on the choice of criteria planners use to attach priorities to sites, as well as the mechanisms employed in the process.

For example, zonation can grow out of existing use patterns, with the end result essentially codifying the already existing segregation of uses. This was largely the case in the original Great Barrier Reef Marine Park zoning plan, especially that of the first zoning done in the Northern Section. Alternatively, zonation may be based on the relative ecological importance of areas within the region and inherent vulnerabilities of different habitats or species. This is the process driving the Belgian biological valuation exercise and is presumably the underlying philosophy of the emerging ocean policy of New Zealand. In stark contrast, zonation may also be based on a kind of conservation-in-reverse process, whereby areas NOT needing as much protection or management as others would be highlighted. Such high-use zones could be 'sacrificial' areas, already so degraded or heavily used that massive amounts of conservation effort would not be cost-effective, or they might be areas determined to be relatively unimportant in an ecological sense.

It is likely that future zoning efforts will use all three of these processes in concert. Existing MPAs and MPA networks will form an important foundation for all zoning plans, regardless of the process or criteria used, because they are de facto precursors of a certain kind of zone. Zoned ocean areas within a region could be administered by a variety of means – by a single overseeing state or federal agency that designs the zoning plans, by a coordinating body that ties together areas variously implemented by different government agencies or by an umbrella framework such as the Biosphere Reserve programme (UNESCO, 1996). The latter has benefits in that local communities become a part of the network, ecologically critical areas are afforded strict protection while less important or less sensitive areas are managed for sustainable use, and the biosphere reserve designation itself carries international prestige (and can be used to leverage funds). For shared coastal and marine resources, regional agreements may prove most effective, especially when such agreements capitalize on better understandings of costs and benefits accruing from shared responsibilities in conserving the marine environment (Kimball, 2001).

Through such large-scale zoning efforts, goals such as biodiversity conservation, conservation of rare and threatened species, maintenance of natural ecosystem functioning at a regional scale, and management of fisheries, recreation, education and research could be addressed in a more coordinated and complementary fashion. The integrated approach inherent in zoning is a natural response to a complex set of ecological processes and environmental problems and is an efficient way to allocate scarce time and resources to combating the issues that parties deem to be most critical. Nations and/or agencies that participate reap the benefits of more effective conservation, while bearing fewer costs by spreading management costs widely and by taking advantage of economies of scale in management training, enforcement, scientific monitoring and the like.

An undeniable trend towards more strategic approaches to marine conservation is apparent in many coastal countries. It is fortunate that these strategies are helping agencies and institutions following their mandates while at the same time overcoming one of the greatest constraints to effective marine conservation: ignoring how they contribute to the big picture beyond their regional, sectoral or agency boundaries (NRC, 2001). An integrated, systematic and hierarchical approach to conservation and sustainable use is needed to allow nations to address various geographic scopes and scales of continental marine conservation problems simultaneously in a more holistic manner (Griffis and Kimball, 1996). By using large marine regions (ecoregions, regional seas, semi-enclosed seas, etc.) as the focus of management rather than individual sites, agencies can come together to address the full spectrum of threats and embark on developing comprehensive zoning plans.

I have attempted to present the case for comprehensive ocean zoning – at the large scale, across all sectors and with the active engagement of all stakeholders in a truly participatory zoning process. Subsequent chapters discuss how ocean zoning differs from the more familiar zoning used by municipalities and regional planners and how zoning can deal with issues of scale, and summarize how steps to institute ocean zoning can be taken. The next two chapters will thus focus on elements of zoning that make up the conceptual framework for the use of this tool at scales relevant to, and necessary for, effective coastal and marine management. Chapter 2 discusses challenges inherent in ocean zoning, while Chapter 3 presents a generic process for developing ocean zoning plans. Chapters 4 to 12 contain short case studies that detail some of the regional planning efforts and ocean zoning schemes that have begun to spring up around the world. The list of case studies is not comprehensive but rather has been selected to illustrate key principles, which are summarized in Chapter 13. The final two chapters then discuss the important issues of implementation and considerations of governance, and summarize what has been learned about ocean zoning to date.

With mounting pressures on marine resources and continued degradation and conflict that arises, this is a dynamic time for marine planning management all around the world. The many ocean zoning initiatives that are in a state of play are among the most dynamic elements of all. Marine spatial planning and ocean zoning are undeniably hot topics today – but we must wait to see whether they will they will turn out to be the solutions that everyone has been waiting for tomorrow.

References

Agardy, T. (2005) 'Global marine policy versus site-level conservation: the mismatch of scales and its implications', invited paper, Marine Ecology Progress Series 300, pp242–248

Agardy, T. (2007) 'How long will we resist ocean zoning? And why?', 'Soapbox', *Sea Technology*, June 2007, p77

Agardy, T. (2009) 'It's time for ocean zoning', *Scientific American Earth 3.0*, Summer 2009, p21

Allison, G. W., J. Lubchenco and M. Carr (1998) 'Marine reserves are necessary but not sufficient for marine conservation'. *Ecological Applications*, vol 8, no 1, S79–S92

Barale, V., N. Schaefer and H. Busschbach (2009) 'Key messages emerging from the ongoing EU debate on Maritime Spatial Planning', available at ec.europa.eu/maritimeaffairs/msp/020709/key_messages_en.pdf

Blaustein, R. (2005) 'Salt in our blood: Is there a philosophy of the oceans?', *Cape Cod Times* 'Forum', 4 September, F-1, F-5

Brierley, A. S. and Michael J. Kingsford (2009) 'Impacts of climate change on marine organisms and ecosystems', *Current Biology*, DOI: 10.1016/j.cub.2009.05.046

Burke, L., Y. Kura, K. Kassem, C. Ravenga, M. Spalding and D. McAllister (2001) *Pilot Assessment of Global Ecosystems: Coastal Ecosystems*, World Resources Institute (WRI), Washington, DC, p94

Christie, P., A. White and P. Deguit (2002) 'Starting point or solution? Community-based marine protected areas in the Philippines', *Journal of Environmental Management*, vol 66, pp441–454

Center for International Earth Science Information Network (CIESIN) (2003) 'Gridded population of the world (GPW)', version 3 beta (online), available at http://quin.unep-wcmc.org/MA/index.cfm (with username and password)

Courtney, F. and J. Wiggin (2003) 'Ocean zoning for the Gulf of Maine: A background paper', www.gulfofmaine.org/council/publications/oceanzoningreport.pdf

Creel, E. (2003) 'Ripple Effects: Population and Coastal Regions'. *Making the Link*, Population Reference Bureau, 8 pp

Crowder, L. and E. Norse (2008) 'Essential ecological insights for ecosystem-based management and marine spatial planning', *Marine Policy*, vol 32, no 5, pp772–778

Crowder, L. B., G. Osherenko, O. R. Young, S. Airame, E. A. Norse, N. Baron (2006) 'Sustainability – resolving mismatches in US ocean governance', *Science*, vol 313, no 5787, pp617–618

Curran, S. R. and T. Agardy (2002) 'Common property systems, migration and coastal ecosystems', *Ambio*, vol 31, no 4, pp303–305

Day, V., R. Paxinos, J. Emmett, A. Wright and M. Goecker (2007) 'The Marine Planning Framework for South Australia: A new ecosystem-based zoning policy for marine management', *Marine Policy*, vol 32, pp535–543

Dayton, P. K., S. F. Thrush, M. T. Agardy, and R. J. Hofman (1995) 'Environmental effects of marine fishing'. *Aquatic Conservation: Marine and Freshwater Ecosystems*, vol 5, no 3, pp205–232

Dayton, P. K., S. Thrush and F. Coleman (2002) 'Ecological effects of fishing in marine ecosystems of the United States', Prepared for the Pew Oceans Commission, Arlington, VA, USA

Ehler, C. and F. Douvere (2007) 'Visions for a Sea Change', *Report of the First International Workshop on Marine Spatial Planning, International Oceanographic Commission and Man and the Biosphere Programme*, IOC Manual and Guides 48, IOCAM Dossier no 4, UNESCO, Paris

Ehler, C. and F. Douvere (2009) Marine spatial planning: A step-by-step approach toward ecosystem-based management, *Intergovernmental Oceanographic Commission and Man and the Biosphere Programme*, IOC Manual and Guides 53, ICAM Dossier no 6, UNESCO, Paris

Epstein, P. R. and J. R. Jenkinson (1993) 'Harmful algal blooms', *Lancet*, vol 342, p1108

European Commission (EC) (2008) *Roadmap for Maritime Spatial Planning: Achieving Common Principles in the EU Brussels*, 25.11.2008 COM(2008) 791 final

FAO Fisheries and Aquaculture Department (2009) 'The state of world fisheries and aquaculture (SOFIA)', text by Jean-François Pulvenis, FAO Fisheries and Aquaculture Department, Rome, available at www.fao.org/fishery/sofia/en, updated 27 February 2009

Griffis, R. B. and K. W. Kimball (1996) 'Ecosystem approaches to coastal and ocean stewardship', *Ecological Applications*, vol 6, no 3, pp708–712

Hall, K. L. et al., (eds) (1992) *Oxford Companion to the Supreme Court of the United States*, pp848–849

Halpern, B. S., K. L. McLeod, A. A. Rosenberg and L. B. Crowder (2008) 'Managing for cumulative impacts in ecosystem-based management through ocean zoning', *Ocean and Coastal Management*, vol 51, pp203–211

Hendrick, D. (2005) 'The new frontier: zoning rules to protect marine resources', *E: The Environmental Magazine*, March–April

Intergovernmental Panel on Climate Change (IPCC) (2003) *Climate Change 2001: The Scientific Basis*, contribution of Working Group I to the Third Assessment Report, J. T. Houghton, Y. Ding, D. J. Griggs, M. Noguer, P. J. van der Linden, X. Dai, K. Maskell, C. A. Johnson (eds), Intergovernmental Panel on Climate Change, Cambridge University Press, Cambridge, UK, 892pp

Kay, R. and J. Alder (2005, 2nd edition) *Coastal Planning and Management*, Taylor and Francis, Abingdon, UK, and New York

Kimball, L. A. (2001) *International Ocean Governance: Using International Law and Organizations to Manage Resources Sustainably*, International Union for Conservation of Nature, Gland, Switzerland, and Cambridge, UK, 124pp

McCay, B. (2008) 'The littoral and the liminal: Challenges to the management of the coastal and marine commons' *MAST 2009*, vol 7, no 1, pp7–30

Millennium Ecosystems Assessment (MEA) (2005) *Ecosystems and Human Well-Being*, Island Press, Washington, DC

MEAM (2008) 'Watershed Management: Putting EBM into Practice, Upstream from the marine environment', *MEAM*, vol 1, no 4, pp1–3

National Research Council (NRC) (1999) *Sustaining Marine Fisheries*, National Academy Press, Washington, DC

National Research Council (NRC) (2001) *Marine Protected Areas: Tools for Sustaining Ocean Ecosystem*, National Academy Press, Washington, DC, 288pp

National Research Council (NRC) (2006) *Increasing Capacity for Stewardship of Oceans and Coasts: A Priority for the 21st Century*, National Academy of Sciences, Washington, DC

Roberts, C. M., G. Branch, R. H. Bustamante, J. C. Castilla, J. Dugan, B. S. Halpern, K. D. Lafferty, H. Leslie, J. Lubchenco, D. McArdle, M. Ruckelshaus, R. R. Warner (2003) 'Application of ecological criteria in selecting marine reserves and developing reserve networks', *Ecological Applications*, vol 13, no 1 (supplement), pp215–228

Rose, J. B., P. R. Epstein, E. K. Lipp, B. H. Sherman, S. M. Bernard and J. A. Patz (2001) 'Climate variability and change in the United States: Potential impacts on water- and foodborne diseases caused by microbiologic agents'. *Environmental Health Perspectives*, vol 109, pp211–221

Sale, P. F. (2008) 'Management of coral reefs: where we have gone wrong and what we can do about it', *Marine Pollution Bulletin*, vol 56, pp805–809

Sanchirico, J. (2009) 'Better defined rights and responsibilities in marine adaptation policy', *Resources for the Future Issue Brief*, pp9–12

Sanchirico, J. N., J. Eagle, S. Palumbi and B.H. Thompson, Jr. (2010) 'Comprehensive planning, dominant-use zones, and user rights: A new era in ocean governance', *Bulletin of Marine Science*, vol 86, no 2, pp273–285

Tibbetts, J. (2002) 'Coastal cities – Living on the edge'. *Environmental Health Perspectives*, vol 110, no 11, A674–A681

UNESCO (1996) *Biosphere Reserves: The Seville Strategy and the Statutory Framework of the World Network*, UNESCO, Paris, 1996, 19pp

UNESCO (2009) ioc3.unesco.org/marinesp

Worm, B., E. B. Barbier, N. Beaumont, J. E. Duffy, C. Folke (2006) 'Impacts of biodiversity loss on ocean ecosystem services', *Science*, vol 314, pp787–790

Worm, B., R. Hilborn, J. K. Baum, T. A. Branch, J. S. Collie, C. Costello, M .J. Fogarty, E. A. Fulton, J. A. Hutchings, S. Jennings, O. P. Jensen, H. K. Lotze, P. M. Mace, T. R. McClanahan, C. Minto, S. R. Palumbi, A. M. Parma, D. Ricard, A. A. Rosenberg, R. Watson and D. Zeller (2009) 'Rebuilding global fisheries', *Science*, vol 325, no 5940, pp578–585

Young, O. R., G. Osherenko, J. Ekstrom, L. B. Crowder, J. Ogden, J. A. Wilson, J. C. Day, F. Douvere, C. N. Ehler, K. L. McLeod, B. S. Halpern, R. Peach (2007) 'Solving the crisis in ocean governance: Place-based management of marine ecosystems', *Environment: Science and Policy for Sustainable Development*, vol 49, no 4, pp20–32

2
Marine Management Challenges: How Ocean Zoning Can Overcome Them

Tundi Agardy

Pelican in the port of Walvis Bay, Namibia

Mismatches between the scale of the problem and its solution

There is a fundamental dilemma in ocean management: the scales at which we can readily practise effective management, and the scales upon which marine ecosystems operate, are wholly different. This is not an esoteric problem; the result of this mismatch of scales is that most of the small-scale management interventions we practise today – regulations regarding a single fishery or establishment of an MPA – do not add up to effective protection across the scales of marine or coastal ecosystems, let alone regions, whole seas or the global ocean (Agardy, 2005).

Successfully addressing environmental problems requires recognition of the problem, mobilization of resources to develop solutions and leadership in driving change. These things are all best accomplished by 'thinking globally, acting locally'. However, environmental problems themselves are rarely local in scale, and piecemeal attempts to address them usually fail. Nowhere is this truer than in the conservation of the marine environment, where open marine ecosystems and the international nature of pollution and other threats dictate a large-scale response. The mismatch between large-scale thinking (embodied in marine policy) and small-scale management actions has serious implications for our ability to reverse the tide of environmental degradation occurring in all the world's oceans.

The problem to date is that operational management – that is, the day to day management actions that safeguard ocean resources and space – occurs at local scales. Despite global and regional agreements and proclamations, effective resource management itself does not happen on the global or regional scale – it occurs bit by bit, as a result of individuals, communities and institutions responding to a particular need at a particular site. Typical marine conservation interventions include marine protected areas, regulations to protect critical habitat of a species and fisheries restrictions pertaining to a threatened stock. The size of these responses is usually far too small to address the bigger (and growing) problems of unsustainable use of resources, indirect degradation of marine ecosystems and large-scale declines in environmental quality, such as those brought about by climate change, since virtually all the world's nearshore areas experience multiple threats that act simultaneously to degrade ecosystems and decrease ecosystem services (Agardy, 2005). And because oceans are the ultimate sink – and the fate of coastal waters is so strongly tied to the condition of coastal lands, rivers and estuaries – successful conservation means addressing not only marine use but land use as well, far up into watersheds.

In contrast, marine policy is generally developed at much larger scales: national, regional and global. These policy initiatives could in theory be broad enough to begin to address holistically complex environmental problems in the oceans. In fact, several international instruments provide impetus for large-scale cooperation, including the United Nations Convention on the Law of the Sea; UN Regional Seas Conventions and Action Plans; Global Programme of Action for the Protection of the Marine Environment from Land-based Activities; Jakarta Mandate on the Conservation and Sustainable Use of Marine and Coastal Biological Diversity; and the RAMSAR Convention on the protection of wetlands (Agardy, 2005). Yet although global treaties and multilateral agreements can bridge some of the gaps that occur between small-scale interventions on the ground and large-scale coastal problems, most of these international instruments have not been effective in reversing environmental degradation (Speth, 2004).

The problem with large-scale policy initiatives is twofold: first, they are by nature too generic to lead to solutions that fit the particular circumstances (environmental, economic, social, political) at a site, and second, they are often unrealistically ambitious or unsupported by financial commitments, and can thus lead nowhere (Wang, 2004). Thus there is a mismatch between what is actually happening and what decision-makers wish was happening.

It is not for lack of want that coastal conservation is failing. Many of the Earth's 123 coastal countries have coastal management plans and legislation, and new governance arrangements and regulations are being developed every year. In 1993 it was estimated that there were 142 coastal management initiatives outside the US and 20 international initiatives, based on an international questionnaire using letters and fax (Sorenson, 1993). By 2000, there was a total of 447 initiatives globally, the result of new initiatives that had started since 1993 and the improved ability to find coastal management initiatives though the use of the internet (Kay and Alder, 2005). The latest survey estimates that there are a total of 698 coastal management initiatives operating in 145 nations or semi-sovereign states, including 76 at the international level (Sorenson, 2002).

Despite a trend towards more strategic approaches to marine conservation, most conservation interventions occur in an ad hoc and opportunistic manner with various agencies and institutions following their mandates devoid of knowing how they fit into and contribute to the big picture beyond their regional, sectoral or agency boundaries (NRC, 2001). An integrated, systematic and hierarchical approach to conservation and sustainable use is needed to allow nations to address various geographic scopes and scales of continental marine conservation problems simultaneously in a more holistic manner (Griffis and Kimball, 1996). A relatively recent movement towards more effective large-scale policy is the coupling of coastal zone management with catchment basin or watershed management, as has occurred under the European Water Framework Directive and projects undertaken under the LOICZ initiative, which have resulted in lower pollutant loads and improved conditions in some estuaries (MEA, 2005). These fully integrated initiatives are best accomplished regionally, with affecting and affected parties taking part in the planning process.

Through such integrated systems, the goals such as biodiversity conservation, conservation of rare and threatened species, maintenance of natural ecosystem functioning at a regional scale, and management of fisheries, recreation, education and research could be addressed in a more coordinated and complementary fashion (Agardy, 2003; Villa et al., 2002). The integrated approach is a natural response to a complex set of ecological processes and environmental problems, and is an efficient way to allocate scarce time and resources to combating critical issues. Through it, top-down, big picture vision can be matched to bottom-up, site-appropriate conservation interventions, and the priorities of international groups that drive attention and money to conservation sites can be harmonized with local priorities and needs.

Comprehensive ocean zoning has the potential to overcome the mismatch and link large-scale marine policy with effective on-the-ground management. For shared coastal and marine resources, regional agreements may prove more effective, especially when such agreements capitalize on better understandings of costs and benefits accruing from shared responsibilities in conserving the marine environment (Kimball, 2001).

Ocean zoning and property rights

Some view the lack of property rights in the ocean as an obstacle to any efforts to apply the land-based principles of zoning to the marine environment. Others see the common property nature of the marine environment as facilitating the use of zoning. Sanchirico (2009) suggests, for instance, that 'it is just the lack of property rights in the ocean that makes the case for ocean zoning so strong'. He and colleagues working on this issue have argued that zoning and the spatial allocation of uses that is part of the process will create 'group property rights' that in turn create stewardship incentives and lead to rationalization of uses, as well as providing new incentives for user groups to organize themselves and become better involved in planning and management (MEAM, 2008a). Zoning can also pave the way towards better management negotiations by endowing all user groups with assets and the flexibility to trade those assets subject to environmental review.

Social scientists typically break property rights into a grouping of operational-level rights, including access (right to enter), withdrawal (right to extract), management (right to regulate use), exclusion (right to deny access) and alienation (right to sell, lease or transfer). Marine protected areas, outer continental shelf leasing and military zones already exist that recognize and limit these kinds of rights. One thing is clear: zoning of communal or common property – such as marine space and resources – is entirely possible by targeting legislation towards use rights.

Dealing with displacement

The prospective costs associated with the establishment of spatial management measures are often held up as a rationale for resisting new ocean zoning or the imposition of stricter regulations in existing zones, such as those that occur in protected zones – portions of an ocean zoning plan where fisheries are limited. Some of the most often cited potential costs are those incurred by fishers through the displacement of fishing effort to new areas or new fisheries.

However, displacement itself is poorly defined and few empirical studies have quantified the impacts of fisheries closure that causes displacement in fisheries. This discussion touches on several different ways of looking at displacement and the evidence that exists for determining the magnitude of displacement impacts, as well as issues of perception and how these play into the reluctance to create and implement zoning plans.

Physical displacement of fishers from one area to other areas

One major concern with ocean zoning and its imposition of place-based regulations on fishing and other extractive activities is the effect that closure of an area to a certain use will have on the user groups and the surrounding environment (Boncouer et al., 2002). Some argue that since reserve designation is rarely done with buyback programmes or other measures that would result in reduced fishing effort, fishing boat displacement will sometimes lead to effort becoming concentrated in smaller areas, causing conflict and ecological harm (Shipp, 2003). Such costs can be avoided by programmes that facilitate alternative livelihoods or provide compensation for lost rights, but usually the funds to support such corollary programmes are not available.

There are potential economic, social and environmental or ecological consequences of displacing efforts, especially if the area closures are not made to be synergistic with other fisheries management measures (Jones, 2006). Putting all the eggs in the marine reserves basket and banking on spillover is much like single species fisheries management, in which the lack of a comprehensive or holistic approach dooms many fisheries management efforts to failure. This is not only because the management intervention of creating a zone closed to fishing (or certain kinds of fishing) may be too narrow in focus, but also because fisheries closures do incur real costs that are too often overlooked, fuelling perceptions or misconceptions that impede future conservation efforts (Agardy et al., 2003).

Unfortunately, though many of the possible costs of fisheries displacement are widely anticipated each time a reserve designation comes up for discussion, there are few analytical studies that actually quantify displacement impacts. The following few paragraphs summarize the types of impacts that are predicted to occur due to displacement.

Economic consequences

Displacement would not be a factor were it not for the increased time and fuel costs of getting to new, presumably farther areas, and the costs associated with learning where new fisheries grounds are or learning new techniques. In theory, these costs could be countered by increased productivity, such that better yields in new areas would compensate fishers for the losses incurred by closing certain fishing grounds (Halpern et al., 2004). One means of potentially increasing productivity is to use strictly protected zones as reserves targeting spawning stock or nursery areas to create enough production that spillover occurs outside reserve boundaries. If stock levels become higher, search costs are lower, all else being equal – a phenomenon known as the stock effect (Sanchirico et al., 2002). Evidence of spillover and stock effect do exist, though some of the evidence has been challenged. Within five years of creation of a network of five small reserves in St Lucia, for instance, fish catches in adjacent waters had increased between 46 per cent and 90 per cent (Roberts et al., 2001). Similarly, increasing numbers of record-size fish in waters adjacent to Florida's Merritt Island National Wildlife Refuge were found after reserve establishment there (Johnson et al., 2000). However, it is difficult to make generalizations based on these studies, since spillover effects vary widely.

In a pan-Europe study of fisheries displacement, six case studies proved to have sufficient data upon which to model the key aspects of the value of fisheries exclusion zones: 1) Gulf of Castellammare, NW Sicily; 2) Western English Channel (sea bass); 3) Firth of Forth, Scotland (Nephrops); 4) Iroise Sea, NW France; 5) Bay of Brest, NW France; and 6) Normand-Breton Gulf, NW France (Pickering, 2003). The results of these modelling exercises are mixed and somewhat ambiguous, but it is clear that the models point to both winners and losers in displaced fisheries. In the Western English Channel, for instance, the UK fleet was predicted to lose some revenues while the western French fleet showed a marginal increase (Whitmarsh, 2003). The modelling study done for European fishery reserves predicted overall decrease in yields, though spawning stock biomass and recruitment were predicted to increase in all case studies (Pickering, 2003).

The situation in California surrounding the Channel Islands reserve process is informative. The Pacific Fisheries Council Scientific and Statistical Committee (SSC) was asked to conduct a technical review of a draft environmental document (DED) prepared by the California Department of Fish and Game for the no-take reserves to be established in the Channel Islands National Marine Sanctuary (CINMS) (SSC, 2002). The SSC was instructed to conduct a general technical review of the DED, keeping in mind any distinctions between the requirements of the California Environmental Quality Act (CEQA) and the National Environmental Policy Act (NEPA). With regard to the effects of effort displacement on fishery resources outside reserves, the DED stated, 'this displacement could cause congestion of effort and a potential negative environmental impact outside MPAs' (SSC, 2002). The SSC noted that the DED provided some information regarding the extent of effort displacement among consumptive recreational users, stating that 63,322 person days of consumptive recreation would be displaced from reserve areas under the proposed project and an additional 14,586 days would be displaced in the federal phase of the project.

Estimated total state-federal displacement under the proposed project comprised 18 per cent of the 437,908 person days of such activity that occur with the Channel Islands National Marine Sanctuary (SSC, 2002). Displacement of commercial fisheries is expressed in the DED in terms of ex-vessel revenues, not fishing effort, suggesting that $3.3 million in harvest would be displaced from reserve areas under the proposed project and an additional $200,000 in the federal phase of the project (SSC, 2002). Total state-federal displacement would account for 16 per cent of the $22.4 million in revenues generated by commercial fishing activities in CINMS (SSC, 2002). While the revenue estimates are categorized by species, the SSC noted that revenues are not necessarily indicative of the amount of effort displaced, since average revenue per unit effort can vary widely among fisheries. While it was not possible to predict precisely what would happen to displaced effort, the Pacific Council felt that fish ticket data could have been used to obtain approximate estimates of the number of trips displaced and the specific CINMS fisheries from which they would be displaced (SSC, 2002). In any case, these predictions of possible displacement costs assume no net benefit provided by reserves.

Spillover has not been demonstrated to the same degree as increased production inside reserves (Ward et al., 2001), but much current scientific research is aimed at determining the extent to which spillover indeed occurs (Gell and Roberts, 2003). Since fishers may soon learn to 'fish the line' – that is, fish close to boundaries of the strictly protected zone in order to take advantage of high catches – they can quickly counteract any spillover effect (McClanahan and Mangi, 2000), though interestingly this has not proven to have occurred in the Georges Bank scallop fishery where fishing the line has been done with great intensity (Murawski et al., 2000). However, there remain reasons to doubt whether reserves can produce the kind of spillover necessary to overcome both costs from physical displacement and perceptions of fishers that they are being unfairly restricted from historic, traditional or most productive fishing grounds. Such congestion externalities also have social and ecological consequences.

Social consequences

The crowding of boats into smaller (and sometimes farther) areas outside zones in which fisheries are restricted has the potential to increase competition and conflict (Shipp, 2003). Often these measures create conflict in fisheries that already are ripe for conflict due to naturally decreasing yields (in other words, those brought about by climate changes) or other issues arising from overcapitalization, decreased environmental health or breakdowns in social institutions or governance mechanisms (Christie, 2004). Paradoxically, the increase in steam time needed to get to new locations outside the closed area is expected to lead to a tendency of fishers to want to invest in greater capital, at a time when many fisheries managers are looking to decrease fishing effort and thus control overcapitalization (Sanchirico et al., 2002). This has not only economic consequences, but also social ones in that fisheries management is less able to do its job, and fishers may become even more unwilling to play by the rules.

There may also be effects that are even more difficult to measure, such as the angst caused by having to learn to fish in new locations. The latter might be an especially important factor in fisheries that have traditionally been located in one place over long periods of time. Safety may also be an issue, if fishers are forced to use existing infrastructure to get to more remote locations or areas that are more vulnerable to storm events, etc. (Sanchirico, 2000). Conflict in new fishing grounds may also erupt as a result of fishers using different gears or coming from different cultures interacting as they never have before.

Ecological consequences

If it is assumed that the new restrictions on fishing brought about by an MPA or reserve are not matched by some other effort reduction, the lack of such exogenous management decisions creates opportunities for overexploitation to continue (Christie et al., 2002). The ecological consequences of overexploitation in turn can be not only stock reduction but also impaired recruitment, selection for smaller sizes of individuals, reduction in genetic diversity and trophic imbalances (Birkeland and Dayton, 2005; Pauly et al., 2002). Clearly effort displaced to depleted stocks would need to be dealt with more restrictively than effort displaced to less-than-fully utilized stocks (Rosenberg, 2003). While potential displacement of effort may also be offset by the potential beneficial effects caused by increased production and spillover from the proposed MPAs (SSC, 2002), many fishers are reluctant to accept this premise on the basis of modelling or the few existing case studies alone (Jones, 2006). Because displacing effort to areas farther offshore or more difficult to reach might lead to investment in greater capital, these ecological impacts could be actually worsened if the protected area measure is not coupled to some form of effort reduction (Sanchirico et al., 2002). However, there is little evidence of this having occurred in real life situations. The lack of information that could provide policy-makers with a starting point from which to evaluate potential effects on fisheries outside reserves and to anticipate what types of specific management actions (if any) might be required to mitigate the effects of displacement precludes a substantive discussion of this issue (SSC, 2002).

A perhaps more troubling ecological effect is the unintended consequence of shifting effort from sink areas to source areas (Crowder et al., 2000). Such a displacement of effort could have dramatic impacts on future recruitment in the wider area, including within the reserve boundaries itself. However, as before, evidence for this actually having occurred is lacking in the published literature. And this potential problem can be avoided with comprehensive ocean zoning – with consideration not only of allowable activities and levels of activity within zones, but in neighbouring zones as well.

Displacement from one fishery into another or into another sector altogether

Zones closed to fishing could conceivably displace fishers not only physically but also into new types of fisheries. Shifting fishing pressure on to new stocks or new species can create increased competition and conflict in the fishery and raise the potential for overexploitation of resources outside the MPA (Sanchirico, 2000). As with the physical displacement of fishers into new areas, the displacement of fishers into new fisheries carries with it the potential for economic, social and ecological costs.

In some cases strict protection zones (or closed areas) will cause fishers to abandon fishing altogether and enter a new employment sector. The most common form of this sort of displacement is from artisanal or small-scale commercial fishing into tourism (Sanchirico et al., 2002). Such displacement is commonly cast in positive terms, as an example of diversifying the economic portfolio of a coastal community or providing alternative livelihoods for people already struggling in highly competitive fisheries. The latter point is particularly germane in areas where artisanal or subsistence fishers have increasing conflicts with the larger scale commercial fishers, as has occurred in the Coromandel Coast of India (Bavinck, 2001) and throughout much of western Africa (Curran and Agardy, 2002). Case studies and syntheses looking at this form of displacement rarely investigate the potential costs of such displacement, which include everything from capital investments in transforming fishing boats into tourism operating vessels to the stigma attached to leaving a traditional livelihood (Christie, 2004). In addition, congestion resulting from tourism development can alter way of life and disturb traditions even for non-fishers (Hoagland et al., 1995).

Perceptions of displacement and how these affect support for marine management

Studies on the ways that spatial management can potentially benefit fisheries have abounded in recent years (Palumbi, 2002). Such analyses are predominantly focused on marine protected areas in tropical seas, but temperate protected zones have also been assessed (Halpern, 2003). In more temperate areas such as the northwestern Atlantic region, Georges Bank closures have been used with much success. In three closed areas there, stocks of haddock, yellowtail flounders and witch flounders increased. Scallops, which had been heavily depleted prior to the closure, rebounded and were 9 to 14 times more abundant than legal-size scallops in fished areas (Murawski et al., 2000).

Despite this evidence and growing advocacy among user groups for spatial management in general, and strictly protected zones in particular, resistance to some forms of management, especially the establishment of large no-take areas, lingers. Spatial management strategies can be viewed as an attempt to police the local community, which may preclude getting the community to support the managed area. This can be especially the case when ocean zoning plans are viewed as being imposed on locals by 'outsiders' (Jones, 2001). Trouble most frequently arises when planners do not recognize that the systems they are managing and studying include people and their sometimes unique cultures. Cultural parameters are especially important to consider in areas having significant populations of indigenous peoples with traditional connections to the marine environment (Crosby et al., 2000; Ward et al., 2001). Attempts to limit access to these resources, especially fishing rights, has the potential to disrupt the socioeconomic stability of coastal communities and result in conflict among user groups with competing interests over the same limited resources. Although the scientific evidence supporting more restrictive access management strategies may be strong, access restrictions are not likely to last long without significant stakeholder support (Agardy, 1997; and see Chapter 13). And because conflict resolution mechanisms can be inadequately incorporated in ocean zoning schemes, the potential for having such perceptions and attitudes derail the zoning plan is great (Christie, 2004; McCreary et al., 2001).

Lack of information, or the perception that not enough attention has been given to the displacement problem, can be a big factor in the success or failure of a marine spatial plan or zoning initiative, at least as far as public or user group acceptance is concerned. Perceptions that zoning regulations unfairly single out a particular user group can also affect compliance and potential for criminal activity and therefore lead to the necessity for increased enforcement investment (Kritzer, 2004). Even when governments acknowledge the potential for displacement and study ways to compensate fishers for lost revenues, as is the case with Australia and New Zealand, fishers remain distrustful because their collective perception is at odds with that of decision-makers.

The real obstacle to understanding displacement is lack of empirical evidence on true costs. While it is true that fishers experience ever-growing costs in most commercial and even some recreational fisheries, many of these costs are related not to having to adjust fishing methods and locations due to closures but rather to increased competition and decreased yields due to declining fish populations, smaller fish sizes and lowered density of fishes within ever-more dispersed populations or stocks. Sometimes the factors behind these declines is natural, such as El Niño-Southern Oscillation (ENSO) events, but more often than not rising costs come from ineffective fisheries management coupled to declining environmental health. What is needed is a controlled experiment in which a healthy fishery experiences a closure and displacement costs are ascertained, all else being equal.

Zoning in dynamic environments

Marine ecosystem boundaries are porous, and most marine systems are both highly dynamic and poorly understood, contributing to significant uncertainty. Many view zoning plans as static and see a disconnect between the problem (managing dynamic ecosystems) and the solution (parcelling ocean space via zoning plans that exist as maps on paper). However, zoning need not be static, and effective zoning plans are never cast in stone.

The act of planning and implementing zones is best done in a periodic manner, occasionally with legal mandates to rezone every few years. In addition, new technologies allow for dynamic zoning – not only in terms of being able to create moving boundaries, but also in terms of technology that improves the ability for users to recognize where those boundaries are in real time. Such dynamic zoning, though complex, does not require high-tech methods for planning, or for effective implementation (that is, management of an area that results in compliance with regulations). Sanchirico quoted in MEAM, 2008b has suggested that zoning which allows trading both within and across zones will introduce private and/or group contracting into the ocean realm that will be a bottom-up way of addressing the fluidity of ocean conditions (MEAM, 2008).

Out beyond the relatively fixed ecosystems of shallow waters, marine ecosystems tend to be highly dynamic, with seasonal and year-to-year changes. But even in these environments ocean zoning can improve conservation of the oceans and management of resources through the use of moving or ephemeral protected areas. For example, pelagic protected areas that have movable boundaries or are ephemeral are now being implemented. Physical oceanography can provide information on how to prioritize spatial management in pelagic environments, with large-scale oceanographic conditions such as frontal systems, upwellings and eddies used to predict concentration of species where spatially explicit management is needed. Such zones could be supplemented with diffuse management in surrounding areas, akin to the use of core areas and buffers, to help protect pelagic species.

Sea turtle conservation provides an example. The National Marine Fisheries Service of the US has been working to identify the temperature regimes where threatened loggerhead turtles congregate across the Pacific Basin – using this information to guide voluntary closures of the Hawaii longline fishery in order to prevent accidental take of this listed species. Thus these closure zones move about in space and time and are dynamic. Dynamic closed areas improve on fixed closed areas because they better match the realities of pelagic species distributions.

Another type of 'special' zone that can be used in dynamic environments is the High Seas protected area designation. High seas are areas beyond the national jurisdiction of countries delineated by their territorial sea or Exclusive Economic Zone (EEZ). An example of such zonation is found in the world's first trilaterally established marine protected area covering both national waters and high seas areas: the Pelagos Sanctuary located in the Ligurian Sea region of the Mediterranean. Much conservation activity is currently focused on identifying key areas to protect on the high seas (see Chapter 9 for an example), yet fewer protected areas exist in the open ocean than in any other ecosystem on Earth.

Countering public misconceptions

Whereas zoning on land is an established practice that has occurred in various forms for centuries, ocean zoning is a relatively new phenomenon and, as such, it has garnered some unease among the public, decision-makers and even some marine planners and managers. Historically, most zoning has been put in place as part of the management framework of multiple-use marine protected areas. Some of the most often-cited examples include the originally mandated zoning and period rezoning of the Great Barrier Reef Marine Park in Australia; the use zones of Netherlands Antillean MPAs such as Bonaire and Saba Marine Parks; the zoning plan developed for the Mafia Island Marine Park in Tanzania; the tri-nationally managed zonation of the Wadden See in northern Europe; and the establishment of 'A', 'B' and 'C' zones in many Mediterranean MPAs, such as those in Italian MPAs, in France's Port Cros Marine Park and elsewhere.

Biosphere reserves, whether on land or in the sea, also utilize zonation by identifying core areas for protection, surrounding those with buffer areas, and casting the whole rest of the reserve as a 'transition zone' or 'zone of cooperation' (Batisse, 1990). UNESCO struggled for years over the question of whether the terrestrial model of biosphere reserves would work in the coastal and marine environment and periodically toyed with the idea of introducing some new concepts in the ocean realm, such as dynamic cores and 'areas of ecological interest' – however, most coastal biosphere reserves are true to the time-tested land form of planning (Agardy, 1997).

It is not meant as a slight on the hard work and dedication of MPA planners to say that zoning within MPAs and Biosphere Reserves is often highly simplistic. The most common zoning patterns are those that restrict all extractive use in certain areas. These closed areas or strict reserves can be created for the purposes of national security, fisheries management, fisheries replenishment or endangered species protection. Sometimes, certain large categories of uses are restricted. A recent example is the designation of large parts of the Arctic Ocean under US jurisdiction as closed to trawl fishing, as done by the North Pacific Regional Fisheries Management Council, and the designation by the General Fisheries Commission of the Mediterranean restricting all bottom trawling in waters greater than 1000m in the Mediterranean.

Thus another major challenge to ocean zoning that is largely absent from land-use or municipal zoning is public misperception and industry apprehension – both commonly resulting from insufficient outreach about the goals of zoning and what zoning entails. For instance, in a workshop held in Nova Scotia on ocean zoning in the Gulf of Maine, a spokesperson for the undersea cable industry remarked that the industry would resist zoning if it resulted in cable zones too narrow to accommodate new installations. Other ocean industries have seen zoning as a tool to deny access; some within the oil and gas industry view zoning as a means for the conservation community to push for blanket prohibitions over wide swaths of the seabed. However, zoning advocates stress the positives of zoning in terms of creating a better climate for private sector investment in both ocean industries and marine conservation, since zoning brings both clarity on rights and some guarantee that rights will be honoured.

Ocean zoning may indeed be inevitable and may prove to be the kind of paradigm-shifting tool of which marine conservationists have long dreamed. Yet not all agree that ocean zoning is a good idea, and there is much reluctance in policy circles for openly embracing the idea and even for using the term 'zoning' in public. This is due in part to philosophical arguments against treating commons property in much the same way we treat private property on land and in part to questions about whether such an inherently complex task as large-scale ocean zoning is feasible.

Many recent articles have spoken to the enormous possibilities presented by ocean zoning (see especially Agardy 2008, 2009; Halpern et al., 2008; Kappel et al., 2009), yet even some cutting-edge conservationists can balk at using the term 'ocean zoning' (e.g., Crowder and Norse, 2008). Similarly, the US Commission on Ocean Policy (2002) fully supported an integrated spatial management but seemed even less willing to insinuate that ocean zoning was a possibility. In a statement on the Joint Initiative of the Joint Oceans Commission, which is a partnership between the US Commission on Ocean Policy and the Pew Oceans Commission that preceded it, it is clear that the zoning concept is embraced, if not explicitly by name: 'development of a coordinated offshore management regime is part of a suite of governance reforms on which the Joint Initiative is now working' (www.jointoceanscommission.org). So it seems that US decision-makers and policy experts are ready to get on the zoning bandwagon, as long as it is not labelled as such.

Another cause for angst in the public and among decision-makers may have its roots in our impatience. Though not inevitable, and not insurmountable, impatience with ocean zoning planning processes can be expected in our increasingly instant-gratification expecting societies. Time horizons for ocean zone planning are expected to be long – even longer than those for MPA planning, which we now recognize as being longer than what the public and decision-makers expect (Agardy, 2005). Impatience with the planning process will doom the effort if there are unrealistic assumptions about how quickly, or how easily, zoning can be created, implemented and adjusted over time. Finally, dynamic marine ecosystems may change so appreciably over relatively short time frames as to make any conventional zoning plans moot (Meir et al., 2004), necessitating the search for dynamic, and truly adaptive, zoning tools.

Besides being a largely unproven methodology, the widespread resistance to ocean zoning may be a natural reaction to diminishing open access and threats to the somewhat mystical appeal of unlimited, unregulated ocean use (Blaustein, 2005). Even highly dedicated marine conservationists may balk at the idea of parcelling the ocean into areas according to their human use values – a common reaction to utilitarian approaches to dealing with nature.

Some writers with what may appear as extremist views assume that zoning state waters, territorial seas and exclusive economic zones is akin to a conservative conspiracy to privatize what should be common property. The privatization of the commons is a controversial topic, particularly the extent to which government can attach property rights or even use rights to entire swaths of ocean. Finally, many planners may fear ocean zoning because it appears to be a fixed construct, wholly unsuitable to the ever-changing marine and coastal environment and the changing human societies supported by it. However, it would be a fallacy to assume that once a zoning plan is developed it is cast in stone. If anything, zoning could pave the way towards two widely claimed goals of the world conservation community: putting ecosystem-based management into practice, and practising adaptive management to increase the efficiency and responsiveness of our marine and coastal management regimes.

Overcoming challenges

Large-scale experiments in ocean zoning that are taking place around the world are helping to overcome both legitimate challenges and misconceptions about what zoning is and how it can be accomplished. These have included, among others, the national efforts of New Zealand to develop its ocean policy. The New Zealand Oceans Policy is meant to achieve integrated and consistent policy for all of the country's waters. The Oceans Policy project started in 2000, under the auspices of the Ministry of Environment, which produced two reports that laid the foundation for comprehensive zoning of New Zealand's national waters: one on priorities in the 200-mile Exclusive Economic Zone (EEZ), the other a gap analysis on ocean management issues not adequately addressed. However, the development of the Oceans Policy was recently put on hold while the government addressed foreshore and seabed issues, and came to terms with scientific uncertainties about priority areas.

The development of New Zealand's Oceans Policy was under way again with the launch of Ocean Survey 20/20, which set out to underpin the development of Oceans Policy and oceans management tools, according to a government proclamation in March 2005 (www.beehive.govt.nz). However, as Ocean Survey 20/20 involves asking the Chief Executives of Land Information New Zealand (LINZ); the Environment, Fisheries and Economic Development Ministries; the Ministry of Research Science & Technology (MORST); Foundation for Research, Science and Technology (FRST); the Defence Force; relevant Crown Research Institute; and the university sector to work together to establish marine information and research priorities over the next 15 years, there is some question about whether the ocean zoning plan will be put on hold for some time to come.

There are also initiatives elsewhere across the globe, as mentioned briefly in Chapter 1. In Belgium, efforts have been under way to define the relative biological values of all parts of the Belgium EEZ (see Derous et al., 2007). In Canada, the government is toying with the idea of ocean zoning as it implements its policies and programmes under the Oceans Act. In the US, the states of Massachusetts, Rhode Island and California have committed to developing zoning plans for state waters (from baseline or coast to 3 nautical miles offshore). Encouraged by the progress by these states, and enabled by visionary leadership within the National Oceanic and Atmospheric Administration, the US federal government committed to developing a marine spatial planning policy for all US waters late in 2009. These and other examples detailed later in this book prove that ocean zoning is feasible and can achieve long sought-after objectives of more effective ocean management.

References

Agardy, T. (1997) *Marine Protected Areas and Ocean Conservation*, R. E. Landes Press, Austin, TX, USA

Agardy, T. (2003) 'Optimal design of individual marine protected areas and MPA systems', in *Aquatic Protected Areas: What Works Best and How do We Know? Proceedings of the World Congress on Aquatic Protected Areas*, Cairns, Australia, August 2002, pp111–119

Agardy, T. (2005) 'Global marine policy versus site-level conservation: the mismatch of scales and its implications', invited paper, *Marine Ecology Progress Series 300*, pp242–248

Agardy, T. S. (2008) 'Casting off the chains that bind us to ineffective ocean management', *Ocean Yearbook*, vol 22, pp1–17

Agardy, T. (2009) 'It's time for ocean zoning', *Scientific American*, vol 19, no 21

Agardy, T., P. Bridgewater, M. P. Crosby, J. Day, P. K. Dayton, R. Kenchington, D. Laffoley, P. McConney, P. A. Murray, J. E. Parks and L. Peau (2003) 'Dangerous targets: Differing perspectives, unresolved issues, and ideological clashes regarding marine protected areas', *Aquatic Conservation: Marine and Freshwater Ecosystems*, vol 13, pp1–15

Batisse, M. (1990) 'Development and implementation of the biosphere reserve concept and its applicability to coastal regions', *Environmental Conservation*, vol 17, no 2, pp111–116

Bavinck, M. (2001) *Conflict and Regulation in the Fisheries of the Coromandel Coast*, Sage Publications, New Dehli

Birkeland, C. and P. K. Dayton (2005) 'The importance in fishery management of leaving the big ones', *Trends in Ecology and Evolution*, vol 20, no 7, pp356–358

Blaustein, R. (2005) 'Salt in our blood: Is there a philosophy of the oceans?' *Cape Cod Times* 'Forum', 4 September, F-1, F-5

Bobko, S. J. and S. A. Berkeley (2004) 'Maturity, ovarian cycle, fecundity, and age-specific parttution of black rockfish (*Sebastes melanops*)', *Fisheries Bulletin*, vol 102, pp418–429

Boncoeur, J., F. Alban, O. Guyader and O. Thébaud (2002) 'Fish, fishers, seals and tourists: Economic consequences of creating a marine reserve in a multi-species, multi-activity context', *Natural Resource Modeling*, special issue: Economic Models of Marine Protected Areas, vol 15, no 4, pp387–412

Christie, P. (2004) 'Marine protected areas as biological successes and social failures in Southeast Asia', *American Fisheries Society Symposium*, no 42, pp155–164

Christie, P., A. White and P. Deguit (2002) 'Starting point or solution? Community-based marine protected areas in the Philippines', *Journal of Environmental Management*, vol 66, pp441–454

Crosby, M. P., R. Bohne and K. Geenen (2000) *Alternative Access Management Strategies for Marine and Coastal Protected Areas: A Reference Manual for their Development and Assessment*, US Man and the Biosphere Program, Washington, DC, 164pp

Crowder, L., S. J. Lyman, W. F. Figueira and J. Priddy (2000) 'Source-sink population dynamics and the problem of siting marine reserves', *Bulletin of Marine Science*, vol 66, no 3, pp799–820

Crowder, L. and E. Norse (2008) 'Essential ecological insights for ecosystem-based management and marine spatial planning', *Marine Policy*, vol 32, no 5, pp772–778

Curran, S. and T. Agardy (2002) 'Common property systems, migration, and coastal ecosystems', *Ambio*, vol 31, no 4, pp303–305

Derous, S., T. Agardy, H. Hillewaert, K. Hostens, G. Jamieson, L. Lieberknecht, J. Mees, I. Moulaert, S. Olenin, D. Paelinckx, M. Rabaut, E. Rachor, J. Roff, E. W. M. Stienen, J. T. Van der Wal, V. Van Lancker, E. Verfaillie, M. Vincx, J. M. Weslawski and S. Degraer (2007) 'A concept for biological valuation in the marine environment', *Oceanologia*, vol 49, no 1, pp99–128

Gell, F. and C. M. Roberts (2003) *The Fishery Effects of Marine Reserves and Fishery Closures*, WWF-US, Washington, DC

Griffis, R. B. and K. W. Kimball (1996) 'Ecosystem approaches to coastal and ocean stewardship', *Ecological Applications*, vol 6, no 3, pp708–712

Halpern, B. S. (2003) 'The impact of marine reserves: do reserves work and does reserve size matter?', *Ecological Applications*, vol 13, no 1, Supplement: S117–137

Halpern, B. S., S. D. Gaines and R. R. Warner (2004) 'Confounding effects of the export of production and the displacement of fishing effort from marine reserves', *Ecological Applications*, vol 14, no 4, pp1248–1256

Halpern, B. S., K. A. Selkoe, C. V. Kappel, F. Micheli, C. D'Agrosa, J. F. Bruno, K. S. Casey, C. Ebert, H. E. Fox, R. Fujita, D. Heinemann, H. S. Lenihan, E. M. P. Madin, M. T. Perry, E. R. Selig, M. Spalding, R. Steneck and R. Watson (2008) 'A global map of human impact on marine ecosystems', *Science*, vol 319, no 5865, pp948–952

Hesseling, R. (1994) 'Displacement: a review of the empirical literature', in R. V. Clarke (ed) *Crime Prevention Studies*, vol 3, Criminal Justice Press, Monsey, NY

Hoagland, P., K. Yashiaki and J. M. Broadus (1995) 'A methodology review of net benefit evaluation for marine reserves', *Environmental Economic Series 027*, World Bank, Washington, DC

Johnson, D. R., J. A. Bohnsack and N. A. Funicelli (2000) 'The effectiveness of an existing estuarine no-take fish sanctuary within the Kennedy Space Center, Florida', *North American Journal of Fishery Management*, vol 19, pp436–453

Jones, P. J. S. (2001) 'Marine protected area strategies: issues, divergences and search for middle ground', *Reviews in Fish Biology and Fisheries*, vol 11, no 3, pp197–216

Jones, P. J. S. (2006) 'Collective action problems posed by no take zones', *Marine Policy*, vol 30, no 2, pp143–156

Kappel, C., B. S. Halpern, R. G. Martone, F. Micheli, K. A. Selkoe (2009) 'In the zone: Comprehensive ocean protection', *Issues in Science and Technology*, Spring 2009, pp33–44

Kay, R. and J. Alder (2005, 2nd edition) *Coastal Planning and Management*, Taylor and Francis, Abingdon, UK, and New York

Kimball, L. A. (2001) *International Ocean Governance: Using International Law and Organizations to Manage Resources Sustainably*, IUCN, Gland, Switzerland, and Cambridge, UK, 124 pp

Kritzer, J. P. (2004) 'Effects of Noncompliance on the Success of Alternative Designs of Marine Protected-Area Networks for Conservation and Fisheries Management', Conservation Biology, vol 18, no 4, pp1021–1031

Millennium Ecosystems Assessment (MEA) (2005) *Ecosystems and Human Well-Being*, Island Press, Washington, DC

MEAM (2008a) 'Comprehensive ocean zoning: Answering questions about this important tool for EBM', *Marine Ecosystems and Management*, vol 2, no 1, pp1–4

MEAM (2008) 'Comprehensive ocean zoning: Answering questions about this powerful tool for EBM', *Marine Ecosystems and Management*, vol 2, no 1, p2

McClanahan, T. R. and S. Mangi (2000) 'Spillover of exploitable fishes from a marine park and its effect on the adjacent fishery', *Ecological Applications*, vol 10, pp1792–1805

McCreary, S., J. Gamman, B. Brooks, L. Whitman, R. Bryson, B. Fuller, A. McInerny and R. Glazer (2001) 'Applying a mediated negotiation framework to integrated coastal zone management', *Coastal Management*, vol 29, pp183–216

Meir, E., S. Andelman and H. P. Possingham (2004) 'Does conservation planning matter in a dynamic and uncertain world?' *Ecology Letters* 7, pp625–622

Murawski, S. A., R. Brown, H. L. Lai, P. J. Rago and L. Hendrickson (2000) 'Large-scale closed areas as a fishery management tool in temperate marine systems: the Georges Bank experience', *Bulletin of Marine Science*, vol 66, no 3, pp775–798

National Research Council (NRC) (2001) *Marine Protected Areas: Tools for Sustaining Ocean Ecosystem*, National Academy Press, Washington, DC, 288pp

Palumbi, S. (2002) 'Marine reserves: A tool for ecosystem management and conservation', *Pew Oceans Commission Science Report*
Pauly, D., V. Christensen, S. Guenette, T. J. Pichter, U. R. Sumaila, C. J. Walters, R. Watson and D. Zeller (2002) 'Towards sustainability in world fisheries', *Nature*, vol 418, pp689–695
Pickering, H. (ed.) (2003) 'The value of exclusion zones as a fisheries management tool: A strategic evaluation and the development of an analytical framework for Europe', *CEMARE Report*, University of Plymouth, UK
Roberts, C. M., J. A. Bohnsack, F. Gell, J. P. Hawkins and R. Goodridge (2001) 'Effects of reserves on adjacent fisheries', *Science*, vol 294, no 5548, pp1920–1923
Rosenberg, A. A. (2003) 'Multiple uses of marine ecosystems', in M. Sinclair and G. Valdimarsson (eds) *Responsible Fisheries in the Marine Ecosystem*, CABI, Wallingford, UK
Sale, R.F., R. K. Cowen, B. S. Danilowicz, G. P. Jones, J. P. Kritzer, K. C. Lindeman, S. Planes, N. V. C. Polunin, G. R. Russ, Y. J. Sadovy and R. S. Steneck (2005) 'Critical science gaps impede use of no-take reserves', *Trends in Ecology and Evolution*, vol 22, no 2, pp74–80
Sanchirico, J. (2009) 'Better defined rights and responsibilities in marine adaptation policy', *Resources for the Future Issue Brief*, pp9–12
Sanchirico, J. N. (2000) 'Marine protected areas as fishery policy: A discussion of potential costs and benefits', *Resources for the Future Discussion Paper*, pp00–23
Sanchirico, J. N., K. A. Cochran and P. M. Emerson (2002) 'Marine protected areas: economic and social implications', *Resources for the Future Discussion Paper*, available at www.environmentaldefense.org/documents/1535_mpas_eco_socio_implic.pdf
Scientific and Statistical Committee of the Pacific Fisheries Council (SSC) (2002) 'Report on the DED', June, available at www.pcouncil.org/reserves/recent/sscreport0602.html
Shipp, R. L. (2003) 'A perspective on marine reserves as a fisheries management tool', *Fisheries*, vol 28, no 12, pp10–20
Sorenson, J. (1993) 'The international proliferation of integrated coastal zone management efforts', *Ocean & Coastal Management*, vol 21, nos 1–3, pp45–80
Sorensen, J. (2002) 'Baseline 2000 background report: The status of integrated coastal management as an international practice (second iteration)', May, available at www.uhi.umb.edu/b2k/baseline2000.pdf
Speth, J. G. (2004) *Red Sky at Morning*, Yale University Press, New Haven
US Commission on Ocean Policy (2004) *An Ocean Blueprint for the 21st Century*, Council of Environmental Quality, Washington, DC
Villa, F., L. Tunesi, and T. Agardy (2001) 'Optimal zoning of a marine protected area: the case of the Asinara National Marine Reserve of Italy', *Conservation Biology*, vol 16, no 2, pp515–526
Wang, H. (2004) 'Ecosystem management and its application to large marine ecosystems: Science, law and politics', *Ocean Development and International Law*, vol 35, pp41–74
Ward, T. J., D. Heinemann and N. Evans (2001) *The Role of Marine Reserves as Fisheries Management Tools: a Review of Concepts, Evidence and International Experience*, Bureau of Rural Sciences, Canberra, Australia
Whitmarsh, D. (2003) 'Synthesis and overview of methodologies used in case studies', in H. Pickering (ed) *The Value of Exclusion Zones as a Fisheries Management Tool*, CEMARE report, University of Plymouth, UK, pp67–74

3
Ocean Zoning Steps

Tundi Agardy

Mangrove conservation site in Tikina Wai, Fiji

Ocean zoning is not a product, it is a process

Ocean zoning is best conceived as a management process utilizing many different tools, as the circumstances warrant and allow. This zoning process is commonly embedded in the wider and much more nebulous management endeavour known as marine spatial planning or maritime spatial planning (the favoured term for integrated marine management in Europe). As discussed in Chapter 1, not all MSP efforts necessarily result in comprehensive ocean zoning, nor does ocean zoning only result from a stated MSP process. I have argued that undertaking the MSP process without committing to zoning misses a great opportunity to improve ocean management and in some cases is just a new way of dressing status quo approaches to new and growing marine management challenges. But let's explore the ocean zoning process itself: where it is generic, and where the process must be tailored to the special circumstances of time and place.

Comprehensive ocean zoning essentially involves a three-part strategy, entailing goal-setting (sometimes called 'visioning'), planning and implementation of management through place-based or area-based regulations. The process is a dynamic, living one – ocean zoning has no logical end point because amendments to management (and thus to zoning plans) must be made continually as ecosystems and human needs change over time. Thus, although many equate zoning with the creation of a map, mapping is only one step in the process, and zoning maps should never be immutably cast in stone.

In some ways, the steps taken to develop and implement a zoning plan can be analogous to those taken in MPA planning. That is, once a vision of comprehensive ocean zoning is created through a participatory process, the target area for zoning must be determined. The region to be zoned can be bounded either by jurisdictional considerations (limits of state or provincial jurisdictions or the outer edge of a nation's Exclusive Economic Zone, for instance), or by ecosystem boundaries, and the areas within the target region can be examined according to their ecological significance, value and use, and condition. Zoning patterns that emerge from a planning process will be based on priorities and compromises between various sorts of objectives for ocean use, as well as feasibility considerations.

It is important to note that while the steps that need to be taken to develop and implement an ocean zoning plan can and should follow a generic process applicable to any marine or coastal management situation, how those steps are taken and which tools are utilized at each step of the way will vary according to circumstances. These circumstances have to do with the nature of the management problem, the social and cultural context, the capacity of local institutions to carry out the management measures and the timeframes available for developing and operationalizing an ocean zoning plan.

The hesitancy to utilize ocean zoning as a management framework that is apparent in many parts of the world may stem from uncertainty about how to begin. Based on in-depth studies of MSP, UNESCO argues that three elements are crucial in beginning a zoning plan (taken from Ehler and Douvere, 2008):

1. Making a plan that can guide the allocation of available resources (people, money, information) within the time required. This includes allocating appropriate amounts of time to key elements of the planning process, e.g., engaging stakeholders, identifying existing conflicts and compatibilities, developing alternative scenarios, identifying management measures and preparing the plan. Developing a work plan should make the best use of available resources
2. Being adaptive. It is a mistake to try to address every issue in the first round of planning; use an open and inclusive stakeholder process to identify spatial management problems that are perceived as 'real' and focus on them first; demonstrate short-term benefits of MSP
3. Focusing on alternative future visions. MSP is about creating a desired future, not simply documenting present conditions and extrapolating current trends

The ocean zoning process thus begins with recognition of the need for zoning, followed by a planning process, a subsequent vetting of the plans and creation of the institutional structures that will allow the plan to be carried out, and implementation that puts the plan in place.

The recognition of need can occur in response to a crisis (e.g., collapse of a fish stock, conflict between resource users, etc.) or catastrophe. Equally it can occur as a result of some systematic evaluation of existing management and its efficacy. Such endogenous evaluation can in turn be initiated by the agencies charged with managing the marine environment, or by outside institutions (government auditor, non-governmental organization, communities or user groups). Recognition of need does not end with the flagging of a problem – optimally it continues with systematic assessments of environmental condition, resource use and management, often referred to as situation analysis. These more specialized appraisals include ecological assessment, risk assessment, rapid rural appraisal and evaluations of the effectiveness of existing management.

The time and effort going into such assessments, and the sophistication of the tools used to determine need, vary from place to place and according to circumstance (institutional capacity). Regardless, the end result is problem-scoping. If it is then determined that the identified problems can be resolved with ocean zoning, the next logical step will be a process, as articulated below.

Articulating the zoning process: visioning, planning and implementing

Visioning and goal-setting

Crucial to the development of any management regime or set of environmental regulations is consideration of the ultimate goal of that management. Too often, management measures are instituted as a kneejerk reaction to a problem, the perception of a problem, or even, strangely enough, when no problem exists, merely because the measure curries favour with the public or a particular set of constituents. (Some campaigning against marine protected area designations in the United States, for example, has labelled no-take marine reserves as 'a management solution in search of a problem'.) Thus, before the planning process begins, a clear understanding of why zoning is being used and to what end needs to be established and communicated.

Goal-setting is especially important when it comes to the marine environment, where there is a perception of commons property, as well as deeply held beliefs in some stakeholder groups that free access should be preserved and that government-mediated management is akin to meddling.

In a recent article in *New Scientist* magazine, the social psychologist Mark van Vugt speaks of ways in which Garrett Hardin's 'Tragedy of the Commons' can be overcome. He articulates four conditions, or '*4i principles*', necessary for successful management of shared environmental resources: information, identity, institutions and incentives. That is, giving the public access to information, making them feel a sense of belonging with a group having common goals, giving them confidence in the ability of institutions to carry out necessary management and identifying incentives to reward non-environmentally destructive behaviour should, he feels, lead to group behaviour that is both altruistic and ecologically sustainable (van Vugt, 2009).

The '*4i principles*' can and should be incorporated in the visioning part of the ocean zoning process. That is, once stakeholders are identified and engaged, they should be encouraged to share their own visions for the goal of ocean zoning by being provided with information about ocean uses, values and condition, as well as being encouraged to discuss issues of identity, institution and incentives.

As stated by Vittorio Barale of the European Union, 'the zoning process itself creates space for an open debate among different maritime sectors, active in a certain area, in order to identify conflicts and means of coexistence between sectors – an objective deemed crucial for ocean management' (Barale et al., 2009).

In order for ocean zoning plans to be achievable – feasible, and at the same time supported by user groups and other stakeholders – the goal-setting or visioning portion of the ocean zoning process should be as participatory as possible. We have learned this time and time again from marine management initiatives around the world, in virtually every social context: when regulations are imposed on users without adequate consultation and participation in planning, management measures are considered unfair and inappropriate, and can be summarily rejected.

That said, participatory process can be very unwieldy and inefficient, and one can well imagine that ocean zoning at large scales and in complicated arenas of maritime use could well be bogged down indefinitely. For this reason, strong leadership is every bit as important as participatory planning.

The next section addresses how zoning is planned, including a discussion of what types of information are needed to do zoning: necessary data, sources of information (whether derived from western social and natural sciences or from traditional knowledge), data analysis, priority-setting and decision-support. But it should be noted that information flow in the planning process is not one-way: information must flow from users and local communities, and from scientists and managers – but it also must flow back to them, in a continually enriching loop.

Planning ocean zoning

The process of planning ocean zoning is itself valuable, not only because of its valuable end-product. Like creating a mathematical model, undertaking the steps of the planning process for zoning provides insights. The planning process forces us to identify stakeholders, recognize connections between use and condition of the system, determine sustainable limits to use and be as strategic as possible. At the core of planning ocean zoning are the tacit assumptions that some areas are more important than others for achieving certain goals, and that this relative importance drives the establishment of spatially explicit rules and regulations.

Planning the spatially explicit management of ocean zoning is about more than creating maps – it is also about meeting the challenges of bounding ecosystems to determine scale and scope of management; assessing ecosystem conditions, threats and drivers; appraising management needs; evaluating trade-offs and choices in order to determine optimal management through a zoning scheme; and planning a monitoring regime to determine efficacy of management and generate, over time, the information necessary to make adjustments to zoning and its corollary regulations.

These challenges are common to any ocean zoning process, regardless of where the efforts are being undertaken and under what environmental, social or political conditions. What differentiates the comprehensive ocean zoning process from standard management is that these steps are strategic and undertaken across large scales that recognize connections, multiple and cumulative impacts, and the need to develop both a holistic vision for marine and coastal management and coordinated management to make it maximally efficient and effective.

The subset of the ocean zoning process that encompasses planning also varies according to circumstance, but can be summarized as having the following ten steps, whether undertaken in a formal or an informal manner (adapted from Agardy, 1997):

1. Identify and involve all stakeholder groups, to the extent practically feasible, in a visioning exercise
2. Set realistic objectives through a participatory process involving relevant stakeholders
3. Study the area (using all applicable science, as well as local knowledge) to determine threats, as well as impediments to realizing objectives
4. Develop outer bounds of the zoning area to reflect objectives and management feasibility
5. Develop a preliminary zoning plan to accommodate different uses – if multiple-use is a goal
6. Amend zoning to reflect user group expectations and needs

7   Formulate a management plan that stipulates permitted uses and levels of use in each zone, in order to address threats (mitigation) and accomplish objectives
8   Develop necessary regulations for each zone and develop incentives to foster voluntary compliance, in order to carry out management
9   Monitor to see if objectives are being met over time
10 Amend management as necessary (adaptive management)

Drawing the appropriate stakeholders into the planning process may be one of the most critical elements of success, but this step is often given scant attention. In the past, participation in the planning of marine management rarely extended beyond just the most obvious resource users (usually fishers) and the government agencies with jurisdiction over the marine area. The result of such limited participation is often backlash against the emergent regulations.

Appropriate stakeholders include both affecting and affected parties (that is, those whose activities, whether on land or at sea, affect ecosystems, and those whose livelihoods or well-being are affected as a result). Such stakeholders can be thought of as being in three general categories: 1) the local community, including civil, non-governmental and labour organizations, 2) the public sector, including central, provincial/state and local government, public service agencies and publicly chartered institutions, and 3) the private sector, including fisheries, aquaculture, energy production and manufacturing industries, waste disposal, tourism, agriculture and forestry (Forst, 2009).

Once an ocean zoning plan is generated in a ten-step process similar to that outlined above, the plan must be vetted by the agencies having responsibility for management, as well as by the public at large. In much of the world, such plans would be published and the public given a comment period to respond. Agencies also must respond within a certain period of time. Subsequent to vetting, there may be a requirement to realign agencies or set up new institutional structures.

Deciding what to protect and what to allow where
Coastal planners and marine resource managers have utilized various tools for identifying priority areas in the past. These approaches vary in information content, scientific rigour and level of technology used, ranging from relatively low-tech participatory planning as often occurs in community-based marine protected area design (Derous et al., 2007) to planning that is driven in part by high-tech decision-support tools such as MARXAN (for example, the zoning of the Great Barrier Marine Park as per Pressey et al., 1997 – see also the next section of this chapter).

Somewhere in the midst of these extremes are methodologies that utilize a variety of tools to optimize site selection through spatial analysis, such as Geographic Information System (GIS)-based multicriteria evaluation (e.g., Villa et al., 2001). The Nature Conservancy (TNC) has recently compiled a set of best practices for Marine Spatial Planning, many of which apply specifically to comprehensive ocean zoning (Beck et al., 2009). The common element of all such approaches is the identification of criteria to discriminate between marine areas and to guide the selection of sites for different uses, or to meet different management objectives. And though the vast majority of these efforts pertain to marine protected area design, there is no reason while such criteria cannot be equally helpful in coastal zone and ocean management more generally.

Marine protected areas can play a key role in such a strategic approach, not because MPAs are a panacea, but rather because MPAs provide a mechanism to overcome two of the biggest obstacles to effective marine conservation. The first of these is the difficulty we have in shedding sectoral management and addressing the full suite of threats to marine ecology in a holistic manner. MPAs provide demonstration models of how to integrate management across all sectors (Villa et al., 2002), and in some cases MPAs demonstrate how to tie ocean management and coastal/watershed management together. The second obstacle that MPAs can help overcome is the management paralysis that arises from the enormous scale of marine problems, their complexity and the strange but pervasive sense that many people have that the oceans are a single homogeneous, fluid environment. MPAs provide an important 'sense of place' to specific habitats and ecological communities, showing that not all parts of the ocean are the same, and so raise the profile and perceived value of specific places in the public's eye. By attaching special importance to specific sites, MPAs not only create opportunities for regulations on use of the area, but also create impetus and political will to address problems that originate outside the area, such as land-based sources of pollution. Individual MPAs are on scales small enough to be tractable, yet a series of MPAs in a strategic network can begin to promote region-wide marine conservation. However, with successful ocean zoning, we must recognize that MPAs will become obsolete – an idea whose time will have come and gone – at least as an end in themselves.

Thus these methodologies have primarily been used to select areas for protection via MPAs. Yet historically the selection of such areas was largely opportunistic or even arbitrary, resulting in decisions that are often hard to defend to the public (Derous et al., 2005). The chance of selecting the areas with the highest intrinsic biological and ecological value through these methods is small (Roberts et al., 2003). Recently, a more Delphic or judgemental approach has been advocated in which an expert panel is consulted to select areas for protection based on expert knowledge. Such a method is relatively straightforward and easily explained, which may indicate why it is still common (Roberts et al., 2003). However, given the sometimes urgent need for site selection, the consultation process is usually too short, the uncertainty surrounding decisions is too high and the information input is too generalized to permit defensible recommendations (Ray, 1999).

The disadvantages of existing methods for assessing the value of marine areas have led to an increasing awareness that a rigorous and objective procedure is needed (Leslie et al., 2003; Possingham et al., 2000). The definition of valuable marine areas should be based on the assessment of areas against a set of objectively chosen criteria, making best use of scientific monitoring data (Mitchell, 1987; Hockey and Branch, 1997; Ray, 1999; Hiscock et al., 2003 –

all cited in Derous et al., 2007). Ecological criteria, which provide a systematic and consistent approach for assessing the biological and ecological importance of marine areas, are also being developed (e.g., Connor et al., 2002). However, the development of such criteria is still in its infancy, and those few ocean zoning initiatives that are currently under way are continuing to use more of a Delphic approach.

Thus a prerequisite to planning the zoning regime is deciding where protection should be maximized. Without information on the relative ecological value of different areas, planners can combine information on intact habitat with use as a proxy for value. Identification of priority sites can then be accomplished with computer algorithms and software such as MARXAN, as has been done in both the Great Barrier Reef Marine Park rezoning (see www.grbmpa.gov.au) and in the Irish Sea Pilot (see www.jncc.gov.uk), or through Delphic methods that utilize expert opinion to develop consensus on key sites. What happens at each site in terms of the actual form of the management regime, however, should be more bottom-up to fit the needs of each particular place (Agardy, 2005). Protected areas, networks and use zones can be imposed by management agencies or organized with partnerships or co-management with indigenous groups and local communities (McCay, 2008).

In exploring the range of methodologies to employ in planning ocean zones, certain assumptions must be made. The first is that enough is known about the ecology of the region to be able to identify key areas needing most protection. In the context of identifying areas as possible no-take fisheries reserves in MPA networks, some have argued that we do not know enough about the life histories of target organisms to adequately site reserves (Sale et al., 2005). A similar argument could of course be made for developing zoning plans. However, ocean zoning is by its nature a coarser exercise, involving the selection of sites for a whole host of criteria, not only a few. While this makes zoning more complex than building MPA networks, it also, paradoxically perhaps, may make it easier. The larger scale and inherently coarser 'grain' of regional ocean zoning allows planners to aggregate information across many different species and hedge bets, as it were. And it should be recognized that adaptive management is at least as important in ocean zoning initiatives as it is in MPAs – perhaps even more important, as environmental conditions, uses and societal needs are more likely to change over time in the larger areas that ocean zoning encompass.

Another assumption that must be made in ocean zoning projects is that benthic-pelagic coupling is strong or, at a minimum, understood well enough to prevent surprise outcomes once management measures are put into place. It is here that the stark differences between land zoning and ocean zoning are most apparent. The terrestrial environment is easier to deal with in that the ecological communities and most of the relevant resources are largely fixed to the substrate. In the sea, the benthic communities are but one target layer, while the water column the benthos is in is another, sometimes wholly different, target. This reality may put people off ocean zoning altogether. However, the situation is not as dire as it might appear: fixed or highly predictable physical features of the ocean environment give significant clues about the relative importance of different parts of the water column, and in some cases, benthic-pelagic coupling is strong.

Using geographic information systems and decision-support tools
There is a sophisticated array of tools that can be employed in the planning stage of zoning. These tools include Geographic Information Systems, Multicriteria Analysis, decision-support software such as MARXAN, MARXAN with Zones and the like, and even Artificial Intelligence systems like ARIES. Later chapters, including specific case studies that outline various methodologies used in ocean zoning, elaborate on some of these tools. However, while the computer-assisted tools used in planning ocean zoning may be complex (and data-intensive), the process of planning is not rocket science and needs to be as straightforward and transparent as possible. This process must use what data are available on ecosystem structure and function, including where important concentrations of ecosystem services occur, as well as on human uses of the ecosystem. First and foremost in creating the maps that will underlie zoning is the objective of identifying areas that are ecologically most critical to ecosystem function, and/or inherently most vulnerable to degradation. These core areas will form the foundation of any spatial management plan, whether it aims to establish marine protected areas, networks of protected areas or comprehensive zoning of the oceans.

The first quantitative methods for systematically identifying 'good' reserve sites were developed in the mid-1970s and used numerical scoring to rank candidate sites in terms of multiple criteria such as species richness, rarity, naturalness and size (Smith and Theberge, 1986). This approach often requires an unreasonably large number of sites to represent all species or other features, because the top-ranked sites frequently contain similar sets of species while missing others; one might need to go far down the ranked list of sites before all species are represented (Williams et al., 2004 cited in Ardron et al., 2008). In contrast to scoring systems, MARXAN allows users to ask the question: what is the minimum number of sites needed to represent all conservation targets (Ardron et al., 2008)?

According to two of the developers of the software program known as MARXAN, used to support the design of marine and terrestrial reserves worldwide, this decision support is the most utilized conservation planning tool; over 60 countries, 1100 users, and 600 organizations use MARXAN to support the design of terrestrial and marine reserves (Ball and Possingham, 2000). Of the approximately 1500 planning efforts using this software, over half have a marine component.

Using MARXAN, conservation planners identify an efficient system of conservation sites that meet a suite of biodiversity targets at a minimal cost. It provides a unique method for designing reserves that is systematic and repeatable. The MARXAN best practices guide (Ardron et al., 2008) produced by PACMARA (see Chapter 11) suggests that the software is best utilized to answer the question: What is the optimum set of planning units needed to meet objectives at minimum cost? (For technical guidance on how to apply MARXAN in developing ocean zoning plans, this manual is invaluable.)

However, as Martin and colleagues point out, MARXAN can also address related questions: What are present gaps in protected areas coverage? How well are protected areas meeting their management targets? How much more area is needed to meet conservation targets? What are the socioeconomic costs of meeting conservation targets? What particular stakeholder groups will be most affected by protected area regulations? What will be the focus of conservation (or management) effort in any area? These are all critical questions in marine management – but as Martin and colleagues so aptly point out, MARXAN has its limitations: it cannot help planners set conservation objectives, determine what levels of protection should be afforded to different sites, engage stakeholders appropriately or address the quality of input data (Martin et al., 2008). Nor can MARXAN help the user determine targets for conservation or management, such as the oft-used target of 20 per cent or 30 per cent of an area that should be set aside as no-take reserve. Such targets represent generalized assumptions and principles, sometimes with limited justification (see Agardy et al., 2003).

Implementing the zoning plan

A disturbing phenomenon in marine management generally is the tendency to invest time and energy in planning without an adequate investment in implementation of the plan. Therefore, part of the ocean zoning planning process must include elaboration of how to implement zoning, with a clear understanding of the costs of implementation, including surveillance, monitoring, enforcement, awareness-raising, education, scientific research, etc.

A diverse matrix of regulations, each pertaining to a particular zone and addressing particular uses, can only be effectively implemented through an integrated approach. Most management entities work in isolation, thus governance that allows cooperation between agencies – and between government and other institutions – is key to being able to implement a zoning plan. Governance arrangements are critical; if institutional structures are insufficient to manage a complex set of regulations associated with a zoning plan, then government or parastatal institutions need to be restructured or newly created (see Chapter 14 for more detail).

Successful implementation does not end with securing adequate budgets for management operations and creating institutional structures that foment integration, cooperation and efficiency in management. Rather, the implementation of zoning is open-ended: zoning efficacy must be constantly re-evaluated and amended if necessary. Lessons learned from ocean zoning within marine protected areas suggest that legally mandated periodic rezoning not only can ensure that the zoning process is effective in meeting management goals but also can build public confidence in and support for the zoning process.

The marine spatial planning process allows certain things to be done well, such as engaging stakeholders from sectors traditionally unaligned or operating independently, attaching values to certain uses of the marine environment, envisioning ways to spatially segregate conflicting uses, and developing scenarios to be able to predict the future consequences of management action. However, doing all of this without having the process lead to an implemented zoning plan misses an opportunity to create truly effective marine management. Subsequent chapters focus in greater detail on certain steps in the zoning process, and the dozen case studies of ocean zoning at various scales provide lessons learned by management agencies attempting to use zoning schemes to meet their management objectives.

Building on what is already there
Quite obviously, no ocean zoning undertaken anywhere in the world starts from a clean slate. Except for the most remote and pristine areas, seas and coasts are already being used by certain user groups, access can be controlled and allocated and some form of management – however inefficient – generally exists. Recognizing what forms of use and management exist is crucial and should form the basis for zoning that integrates across sectors and across wider sets of ecosystems.

Zonation can grow out of existing use patterns, with the end result essentially codifying the already existing segregation of uses. This was largely the case in the original Great Barrier Reef Marine Park zoning plan (see Chapter 4), especially that of the first zoning done in the Northern Section (Agardy, 1997). Alternatively, zonation may be based on the relative ecological importance of areas within the region and the inherent vulnerabilities of different habitats or species. This is the process that drove the Belgian biological valuation exercise (see Chapter 7) and presumably underlies the emerging Oceans Policy of New Zealand (Chapter 5). In stark contrast, zonation may also be based on a kind of conservation-in-reverse process, whereby areas *not* needing as much protection or management as others would be highlighted. Such high-use zones could be 'sacrificial' areas, already so degraded or heavily used that massive amounts of conservation effort would not be cost-effective, or they might be areas determined to be relatively unimportant in an ecological sense.

It is likely that future zoning efforts will use all three of these processes in concert. Existing MPAs and MPA networks will form an important foundation for all zoning plans, regardless of the process or criteria used, because they are precursors of a certain kind of zone. This brings us to the final section of this chapter: using MPAs as a starting point for ocean zoning.

MPAs as a starting point

In striving for management that is effective in addressing ever-increasing marine problems, marine managers often reach for marine protected areas and marine protected area networks among the first tools. Yet although MPAs are likely to be necessary in the development of effective ecosystem-based management, they are unlikely to be sufficient in reaching that overarching goal. (Christie et al. [2002] refers to MPAs as the 'starting point, not solution', in reference to marine management in the Philippines.) Nonetheless, using core protected areas to spatially 'ground' marine management is a sound strategy that many marine planners and management agencies employ in developing widerscale, more integrated marine management. Because zoning plans don't exist in a vacuum, it makes sense to investigate how well-planned MPAs are designed, and build upon existing protection regimes to develop comprehensive ocean zoning.

Marine protected areas can thus be a useful starting point in comprehensive ocean zoning, whether the zoning spans coastal seas, ocean basins or large marine ecosystems (see MEAM, 2008). If MPAs have been strategically designed to protect what is most critical or vulnerable, then a foundation for strategic planning of the wider areas already exists. When using MPAs as a starting point, planning can proceed by 1) looking for gaps in protected area coverage, 2) evaluating the effectiveness of existing MPAs, and 3) determining what uses and levels of use can be permitted in the matrix surrounding protected areas. However, it should be noted that not all protected areas are strategically planned – they sometimes are designated in an ad hoc fashion, or they may be instituted to serve purposes other than management for sustainability.

MPA benefits

With increasing recognition of the threats to the marine and coastal environment, a wide variety of MPAs and related policy frameworks has been developed to conserve and sustainably use coastal and marine resources and ecosystems. Overarching goals for MPAs can be thought of as related to conservation or to sustainable use, though in many MPAs the goal is to practise both of these within a circumscribed spatial scale. MPAs can help achieve many objectives, including:

1. Biodiversity conservation
2. Conservation of rare and restricted-range species
3. Maintenance and/or restoration of a natural ecosystem functioning at local and regional scales
4. Conservation of areas vital for vulnerable life stages
5. Reducing or minimizing user conflict
6. Managing fisheries (using reserves to sustain or enhance yields, restore or rebuild stocks of overexploited species, and provide insurance against management failures)
7. Recreation
8. Education
9. Research
10. Fulfilling aesthetic needs

Marine protected areas are also established as research sites to further understanding of marine ecosystems and human impacts on them. Adaptive management, in which monitoring of uses and compliance with regulations and its effect on ecosystem condition, is used to derive information for amending management (e.g., moving boundaries of protection, changing regulations within management areas, adding or removing catch limits, etc.) so it remains effective over time and changing conditions (both environmental and societal). MPAs can be central to the sort of adaptive management that can make zoning schemes effective by providing the control sites and experimental conditions for understanding the effects of management.

Management of MPAs for marine species conservation is complex, requiring sometimes even some knowledge of terrestrial environments as well as marine ones. MEAM (2008) describes Jon Day of the Great Barrier Reef Marine Park Authority illustrating this point by referring to the plight of the green sea turtle on the Great Barrier Reef. These turtles lay their eggs on the mainland or islands outside the Great Barrier Reef Marine Park; once hatched the young turtles move into nearshore marine areas in the park, feeding on seaweed and seagrasses. They then migrate thousands of kilometres in open sea to other countries, before returning to Australia to the same stretch of beach where they were born. This means that effective conservation of this species alone needs to consider local, provincial, national and international jurisdictions (both in the open sea and within various countries).

Out beyond the relatively fixed ecosystems of marshes, mangroves, rocky shorelines, seagrasses and coral reefs, marine ecosystems tend to be highly dynamic, with seasonal and year-to-year changes. MPAs can help provide the foundation for marine management of dynamic elements of ecosystems in at least three ways: 1) through the use of moving or ephemeral protected areas, 2) through the use of fixed MPAs on the high seas, and 3) through the use of MPAs that are established specifically to promote research on the ecology of understudied ecosystems and the efficacy of management in those areas (Hyrenbach et al., 2000).

Pelagic protected areas that have movable boundaries or are ephemeral are being explored by planners and researchers alike. Large-scale oceanographic conditions such as frontal systems, upwellings and eddies can be used to predict concentration of species where spatially explicit management is needed. Supplementing such dynamic MPAs with diffuse management in surrounding areas, akin to the use of core areas and buffers in a zoning scheme, would move us closer to true ecosystem-based management (MEAM, 2008).

The US National Marine Fisheries Service (NMFs) has been working with colleagues to identify the temperature regimes where threatened loggerhead turtles congregate across the Pacific Basin – using this information to guide voluntary closures of the Hawaii longline fishery to prevent accidental take of this listed species. NMFS researchers feel that dynamic closed areas improve on fixed closed areas, because they better match the realities of (pelagic) species distributions (MEAM, 2008; for greater detail on this innovative programme to establish dynamic MPAs, see www.int-res.com/articles/esr2008/5/n005p267. pdf). Similar fishery closures are being designed for other pelagic marine species, such as around bluefin tuna spawning areas in the Indo-Pacific.

Since large marine ecosystems are rarely confined to the waters of coastal nations, extending management to the high seas (beyond national jurisdictions) is a logical element of effective management, and can be invaluable in forming the basis for zoning plans. High seas MPAs are much less common than MPAs established in nearshore waters, and the first high seas MPA – the Pelagos Sanctuary located in the Ligurian Sea region of the Mediterranean – was established only a few years ago (Notarbartolo di Sciara et al., 2007). This tri-nationally planned protected area encompasses both national waters and high seas areas, and its establishment has spurred interest in identifying further areas to protect on the high seas – in the Mediterranean and beyond.

It should also be noted that MPAs are useful not only for in situ conservation, but also in generating support for wider scale marine management. Protected areas can 'put a face on a place' – allowing people to better relate to vast and seemingly amorphous tracts of ocean. Protected areas also provide engagement in planning, and can be used to bring people into planning and management processes. Protected areas can lift the discussion on management out of the constrained realm of scientists and management agencies to broader discussions that involve user groups and affected communities. MPAs can raise awareness about marine conservation, can be used in education about the marine environment and can provide the framework to engage and empower stakeholders. Given the challenge of practising marine management at the scales appropriate to large marine systems, this motivation is essential.

MPA networks

The use of individual marine protected areas has been flourishing. In recent years, networks of marine protected areas have been planned to achieve what single protected areas could not: to protect wider ecosystems by strategically placing protections in key locales. These networks began to emerge when planners realized that few marine protected areas were meeting broad-scale conservation objectives, and that an ad hoc, one-off approach would not lead to effective large-scale conservation.

MPA networks are a way to plan MPAs strategically so that the whole is greater than the sum of its parts. Various sorts of MPA networks and systems include representative systems, systems of MPAs designed to protect a single or suite of target species, MPA networks designed to protect sources and sinks of larval recruits for key species (such as corals or commercially important fish), and MPA networks aimed at protecting critical, linked habitats in a wider region – such as those that span coastal wetlands and soft- and hard-bottom communities offshore.

There is an obvious place for strategically developed marine protected area networks in the marine management toolbox (Roberts et al., 2001). A system or network that links these areas has a dual nature: connecting physical sites deemed ecologically critical (ecological networks), and linking people and institutions in order to make effective conservation possible (human networks) (Agardy and Wilkinson, 2004). Networks or systems of marine protected areas have great advantages in that they spread the costs of habitat protection across a wider array of user groups and communities while providing benefits to all, and they help to overcome the mismatch of scale (Agardy, 2003).

The last decade has witnessed a surge in MPA networks as a way to meet national goals for marine management. For instance, the government of New Zealand is investing in developing MPA networks through its Department of Conservation, within the Ministry of Fisheries. In 2008 the government of New Zealand issued a set of guidelines detailing the planning process for MPA networks (Department of Conservation, 2009).

Protected area networks that abate threats to a particular species in a particular place, or that conserve critical habitats for key species, can do more than provide species protection when the species in question is an umbrella species (organisms having broad ecological requirements that are shared by many other species), keystone species (organisms playing a pivotal role in ecosystem dynamics), or even a flagship species (charismatic organisms that draw attention to conservation issues). Conservation measures aimed at mitigating threats to such species result in protection for whole communities of organisms, and indeed whole ecosystems. In light of this, investments in conservation of such species can be seen as investments in maintaining overall marine biodiversity and ecosystem functioning.

A rational approach to conserving or restoring marine populations is to use strictly protected areas to safeguard critical habitats such as feeding areas, breeding areas and resting or staging grounds. But since these areas are often separated by hundreds of kilometres, the areas between critical habitats are also of concern. Identifying and then protecting migration corridors is one technique to ensure the links between the critical habitats remain unbroken. Another is to broaden management to a regional scale, utilizing comprehensive ocean zoning, to help address threats. Regulations within MPAs, within networks and in the buffer areas in-between should be tailored to those threats for MPAs to be able to meet their species conservation targets. This, essentially, is what ocean zoning achieves: it allows protection of special species, vulnerable areas or valuable ecosystem services through place-based conservation, while also tackling the condition of the wider ecosystems in which these islands of protection sit.

Thus, planners and managers have increasingly acknowledged the importance of looking at areas beyond protected area boundaries. David Johnson, Executive Secretary of the intergovernmental OSPAR Commission, has suggested that while core zones that form the basis of MPAs can conserve critical areas, it may be every bit as important to focus attention on the buffer zones around protected areas. These areas must be carefully managed, since their condition will likely have a profound influence on the viability of these islands of protection (MEAM, 2008).

This is especially true because although MPAs and MPA networks are fast becoming the conservation tool of choice for dealing with habitat loss, and are increasingly being used to address fisheries problems and to involve local communities and user groups in management of marine areas, such areas are usually far too small to be effective in addressing the complex suite of problems that most marine areas face (Agardy, 2005). This is especially true when planners and conservation groups ignore the context in which these islands of protection sit (Allison et al., 1998). In addition, generic but simplistic models of marine protected areas can be counterproductive, in that applying a cookie-cutter approach can lead to management failure and distrust (Agardy et al., 2003).

A real quantum leap in conservation effectiveness can occur when planners scale up from MPA networks and corridor concepts to full-scale ocean zoning. Ocean zoning provides many benefits over smaller scale interventions: it can help overcome the shortcomings of MPAs and MPA networks in moving us towards sustainability; it is based on a recognition of the relative ecological importance and environmental vulnerabilities of different areas; it allows harmonization with terrestrial land-use planning; and it can help better articulate private sector roles and responsibilities, as well as maximize private sector investment by capitalizing on free market principles.

References

Agardy, T. (1997) *Marine Protected Areas and Ocean Conservation*, R. E. Landes Press, Austin, TX, USA

Agardy, T. (2003) 'Optimal design of individual marine protected areas and MPA systems', in *Aquatic Protected Areas: What Works Best and How do We Know? Proceedings of the World Congress on Aquatic Protected Areas*, Cairns, Australia, August 2002, pp111–119

Agardy, T. (2005) 'Global marine conservation policy versus site level implementation: the mismatch of scale and its implications', in H. I. Browman and K. I. Stergiou (eds), 'Politics and socio-economics of ecosystem-based management of marine resources', Journal Special Issue, *Marine Ecology Progress Series* 300, pp241–296

Agardy, T., P. Bridgewater, M. P. Crosby, J. Day, P. K. Dayton, R. Kenchington, D. Laffoley, P. McConney, P. A. Murray, J. E. Parks and L. Peau (2003) 'Dangerous targets: Differing perspectives, unresolved issues, and ideological clashes regarding marine protected areas', *Aquatic Conservation: Marine and Freshwater Ecosystems*, vol 13, pp1–15

Agardy, T. and T. Wilkinson (2004) 'Towards a North American network of marine protected areas', *Proceedings of the Conference on Science and Management of Protected Areas*, 11–15 May 2003, Victoria, BC, Canada

Allison, G. W., J. Lubchenco and M. Carr (1998) 'Marine reserves are necessary but not sufficient for marine conservation', *Ecological Applications*, vol 8, no 1, S79–S92

Ardron, J., H. P. Possingham and C. J. Klein (2008) *Marxan Good Practices Handbook*, PACMARA, Pacific Marine Analysis and Research Association, Vancouver, BC, Canada, available at www.pacmara.org

Ball, I. R. and Possingham, H. P. (2000) *Marxan (V1.8.2): Marine Reserve Design Using Spatially Explicit Annealing, a Manual*, University of Queensland, Brisbane, Australia

Barale, V., N. Schaefer and H. Busschbach (2009) 'Key messages emerging from the ongoing EU debate on Maritime Spatial Planning', available at europa.eu/maritimeaffairs/msp/020709/key_messages_en.pdf

Beck, M. W., Z. Ferdana, J. Kachmar, K. K. Morrison, P. Taylor and others (2009) *Best Practices for Marine Spatial Planning*, The Nature Conservancy, Arlington, VA, USA

Christie P., A. White and P. Deguit (2002) 'Starting point or solution? Community-based marine protected areas in the Philippines', *Journal of Environmental Management*, vol 66, pp441–454

Connor, D.W., J. Breen, A. Champion, P. M. Gilliland, D. Huggett, C. Johnston, D. Laffoley, L. Lieberknecht, C. Lumb, K. Ramsay, M. Shardlow (2002) 'Rationale and criteria for the identification of nationally important marine nature conservation features and areas in the UK Version 02.11', Joint Nature Conservation Committee for the DEFRA Working Group on the Review of Marine Nature Conservation Working Paper, Peterborough: 23 pp

Department of Conservation (New Zealand) (2009) www.doc.govt.nz/getting-involved/consultations/results/marine-protected-areas-classification-protection-and-guidelines/

Derous, S., T. Agardy, H. Hillewaert, K. Hostens, G. Jamieson, L. Lieberknecht, J. Mees, I. Moulaert, S. Olenin, D. Paelinckx, M. Rabaut, E. Rachor, J. Roff, E. W. M. Stienen, J. T. van der Wal, V. Van Lancker, E. Verfaillie, M. Vincx, J. M. Weslawski and S. Degraer (2007) 'A concept for biological valuation in the marine environment', *Oceanologia*, vol 49, no 1, pp99–128

Ehler, C. and F. Douvere (2008) 'How do you begin an ocean zoning process?', *MEAM*, September–November 2008, p3

Fairweather, P. G. and S. E. McNeill (1993) 'Ecological and other scientific imperatives for marine and estuarine conservation', in 'Protection of marine and estuarine environments – a challenge for Australians', A. M. Ivanovici, D. Tarte, M. Olsen (eds) *Proceedings of the Fourth Fenner Conference*, Canberra, 9–11 October 1991, *Occasional Paper No 4*, Department of Environment, Sport and Territories, Canberra, pp39–49

Forst, M. F. (2009) 'The convergence of integrated coastal management and the ecosystems approach', *Ocean and Coastal Management*, vol 52, no 6, pp296–306

Hiscock, K., M. Elliott, D. Laffoley and S. Rogers (2003) 'Data use and information creation: Challenges for marine scientists and for managers', *Marine Pollution Bulletin*, vol 46, no 5, pp534–541

Hockey, P. A. R. and G. M. Branch (1997) 'Criteria, objectives and methodology for evaluating marine protected areas in South Africa', *South African Journal of Marine Science*, vol 18, pp369–383

Hyrenbach, K. D., K. A. Forney and P. K. Dayton (2000) 'Marine protected areas and ocean basin management', *Aquatic Conservation: Marine and Freshwater Ecosystems*, vol 10, pp437–458

Leslie, H., M. Ruckelshaus, I. R. Ball, S. Andelman and H. P. Possingham (2003) 'Using siting algorithms in the design of marine reserve networks', *Ecological Applications*, vol 13, no 1, supplement 2003, S185–S198

Martin, T. G., J. L. Smith, K. Royle and F. Heuttmann (2008) 'Is Marxan the right tool? Chapter 2', in J. Ardron (ed.) *Marxan Good Practices Handbook*, Pacific Marine Analysis and Research Association, Vancouver, BC, Canada, pp20–25

McCay, B. (2008) 'The littoral and the liminal: Challenges to the management of the coastal and marine commons', *MAST 2009*, vol 7, no 1, pp7–30

MEAM (2008) 'Comprehensive ocean zoning: Answering questions about this important tool for EBM', *Marine Ecosystems and Management*, vol 2, no 1, pp1–4

Mitchell, R. (1987) *Conservation of Marine Benthic Biocenoses in the North Sea and the Baltic: A Framework for the Establishment of a European Network of Marine Protected Areas in the North Sea and the Baltic*, Council of Europe, European Committee for the Conservation of Nature and Natural Resources, Strasbourg

Notarbartolo di Sciara, G., T. Agardy, D. Hyrenbach, T. Scovazzi, P. Van Klaveren (2007) 'The Pelagos Sanctuary for Mediterranean marine mammals', *Aquatic Conservation: Marine and Freshwater Ecosystems*, vol 18, pp367–391, DOI: 10.1002/aqc.855

Possingham H. P., I. R. Ball and S. Andelman (2000) 'Mathematical methods for identifying representative reserve networks', in S. Ferson and M. Burgman (eds) *Quantitative Methods for Conservation Biology*, Springer-Verlag, Düsseldorf, Germany

Pressey, R. L., H. P. Possingham and J. R. Day (1997) 'Effectiveness of alternative heuristic algorithms for identifying indicative minimum requirements for conservation reserves', *Biological Conservation*, vol 80, pp207–219

Ray, C. C. (1999) 'Coastal-marine protected areas: agonies of choice', *Aquatic Conservation: Marine Freshwater Ecosystems*, vol 9, pp607–614

Roberts, C. M., G. Branch, R. H. Bustamante, J. C. Castilla, J. Dugan, B. S. Halpern, K. D. Lafferty, H. Leslie, J. Lubchenco, D. McArdle, M. Ruckelshaus and R. R. Warner (2003) 'Application of ecological criteria in selecting marine reserves and developing reserve networks', *Ecological Applications*, vol 13, no 1, S215–228

Roberts, C. M., B. Halpern, S. R. Palumbi and R. R. Warner (2001) 'Designing marine reserve networks: Why small, isolated protected areas are not enough', *Conservation Biology in Practice*, vol 2, no 3, pp12–19

Sale, R. F., R. K. Cowen, B. S. Danilowicz, G. P. Jones, J. P. Kritzer, K. C. Lindeman, S. Planes, N. V. C. Polunin, G. R. Russ, Y. J. Sadovy and R. S. Steneck (2005) 'Critical science gaps impede use of no-take reserves', *Trends in Ecology and Evolution*, vol 22, no 2, pp74–80

Smith, P. G. R. and J. B. Theberge (1986) 'A review of criteria for evaluating natural areas', *Environmental Management*, vol 10, pp715–34

van Vugt, M. (2009) 'The triumph of the commons', *New Scientist*, 22 Aug 2009, pp40–43

Verreet, G. (2009) 'Ecosystem based approach, MSP and link with the Marine Strategy', available at eurlex.europa.eu/LexUriServ/LexUriServ.do?uri=CELEX:DKEY=483715:EN:NOT

Villa, F., L. Tunesi and T. Agardy (2001) 'Optimal zoning of a marine protected area: the case of the Asinara National Marine Reserve of Italy', *Conservation Biology*, vol 16, no 2, pp515–526

Williams, J. C., C. S. ReVelle and S. A. Levin (2004) 'Using mathematical optimization models to design nature reserves', *Frontiers in Ecology and the Environment*, vol 2, no 2, pp98–105

4
Zoning within the Great Barrier Reef Marine Park (Australia)

Tundi Agardy

Reef flat in the Great Barrier Reef Marine Park

Great Barrier Reef Marine Park

The Great Barrier Reef Marine Park (GBRMP) of Australia is an iconic marine park that may well provide the best example of large-scale ocean zoning in existence today. Located off the eastern Australian coast and paralleling the coastline of the state of Queensland, the GBRMP spans 344,400km^2 and runs over 2000km in length (see Colour Plate 4.1). When it was established in 1975, it had the distinction of being the largest marine protected area in the world (until several recent MPA designations usurped its status). The zoning schemes and associated management approaches used within this multiple-use park are often held up as the best example of the integration of management using ocean zoning. However, the GBRMP's large size undeniably creates challenges for park-wide operational management; for this reason, the GBRMP was initially divided into three sections that each had the capacity to manage or regulate impacts through zoning plans.

When the initial legislation for the GBRMP was enacted in 1975, it established the planning process for the GBRMP, specified that certain aspects of public participation in planning were mandatory and implied that zoning plans were to be one of the key management tools. The initial legislation did include a specific section requiring the preparation of a zoning plan for any area declared to be part of the GBRMP; another section required the preparation of a zoning plan to have regard to the following objectives:

- the conservation of the Great Barrier Reef
- the regulation of the use of the Marine Park so as to protect the Great Barrier Reef while allowing the reasonable use of the Great Barrier Reef Region
- the regulation of activities that exploit the resources of the Great Barrier Reef Region so as to minimize the effect of those activities on the Great Barrier Reef
- the reservation of some areas of the Great Barrier Reef for its appreciation and enjoyment by the public
- the preservation of some areas of the Great Barrier Reef in its natural state undisturbed by man except for the purposes of scientific research

The first zoning plans were progressively developed for parts of the Marine Park in the early 1980s, but it was not until 1988 – fifteen years after the Marine Park was declared – that the entire area was zoned (see Colour Plate 4.1). From 1988 until mid-2004, less than 5 per cent of the entire GBR was zoned in highly protected 'no-take' zones.

While the initial zoning scheme sought to reduce conflict over fishing, issues, uses and community expectations all changed over time, necessitating a rezoning effort. This was because the original zoning was biased towards coral reefs, whereas other significant parts of the ecosystem were either inadequately protected or not protected at all. The original zoning of the Great Barrier Reef Marine Park is akin to the type of land-use zoning that has been practised for centuries. Kay and Alder state that zoning originally grew from 'nuisance' crisis management in response to health, sanitation and transportation problems (Kay and Alder, 2005). They quote Hall et al., 1993: 'In Britain, as elsewhere, town planning had grown up as a local system of zoning control designed to avoid bad neighbour problems and to hold down municipal costs' (Kay and Alder, 2005). The natural extension of this sort of planning was to apply zoning as a tool to control shipping to avoid collisions at sea. Eventually, MPA and coastal planners began to see the fisheries industry as a potentially 'bad neighbour', and the drive to limit fishing (especially large-scale commercial fishing) to certain areas created the need for more comprehensive ocean zoning. Thus, the initial GBR zoning scheme (developed in 1981) was focused primarily on extractive uses, and the total amount of area where extractive activities were prohibited was originally quite small (around 4 per cent) until the Representative Areas Program began in the late 1990s.

The Great Barrier Reef Marine Park in Australia is without question the most ambitiously planned MPA in the world. Its vast expanse reaches from far north Queensland to the southern margin of the reef some 1600km to the south. The 350,000km^2 marine park covers the largest single collection of coral reefs and associated habitats in the world.

The GBRMP was established in 1975 through an Act of the Australian Parliament. An independent statutory body called the Great Barrier Reef Marine Park Authority (GBRMPA) oversees planning and management and has its main headquarters and majority of staff in Townsville. Field management is undertaken by a variety of federal and state (Queensland) agencies spread along the length of the GBRMP, but all permitting, enforcement, research and zoning activities are coordinated by the GBRMPA.

The primary objective the forward looking GBRMP initially aimed to achieve through its multiple use zoning plan was to accommodate anticipated growth in coastal and marine tourism while at the same time avoiding conflicts with other economic sectors. The GBRMPA believes 'any use of the GBR should not threaten its existing ecological characteristics and processes' and adopts as the primary goal of the GBRMP 'the long term protection, ecologically sustainable use, understanding, and enjoyment of the Great Barrier Reef' (GBRMPA, 2009a). Tourism and related development revenues exceed AU$5 billion annually (GBRMPA, 2009b) and tourism accommodation is clearly a main objective – one that some have argued was initially promoted at the expense of conservation in the original zoning plans.

Across the entire span of the GBRMP, different areas are designated as being of one of eight different possible zones: 1) General Use, 2) Habitat Protection, 3) Conservation Park, 4) Buffer, 5) Scientific Research, 6) National Park, 7) Preservation or 8) Commonwealth Islands (GBRMPA, 2003). These zones each have very specific objectives behind their designation (see Table 4.1).

Activities allowed 'as-of-right' and those activities that require a permit within each zone are listed in Table 4.2. They include fishing and other kinds of natural resource harvesting, non-extractive uses such as diving and photography, shipping, and scientific research.

Table 4.1 *Types of Zones (with colours to correspond to zoning map in Plate 4.1) in the GBRMP and objectives for each (Source: GBRMPA)*

Zone	Objectives
General Use Zone (light blue) Comparable IUCN category – VI	To provide for the conservation of areas of the Marine Park while providing opportunities for reasonable use.
Habitat Protection Zone (dark blue) Comparable IUCN category – VI	To provide for the conservation of areas of the Marine Park through the protection and management of sensitive habitats, generally free from potentially damaging activities, and, subject to the above, to provide opportunities for reasonable use.
Conservation Park Zone (yellow) Comparable IUCN category – IV	To provide for the conservation of areas of the Marine Park, and, subject to the above, to provide opportunities for reasonable use and enjoyment, including limited extractive use.
Buffer Zone (olive) Comparable IUCN category – IV	To provide for the protection of the natural integrity and values of areas of the Marine Park, generally free from extractive activities, and, subject to the above, to provide opportunities for: (a) certain activities, including the presentation of the values of the Marine Park, to be undertaken in relatively undisturbed areas and (b) trolling for pelagic species.
Scientific Research Zone (orange) Comparable IUCN category – IA	To provide for the protection of the natural integrity and values of areas of the Marine Park, generally free from extractive activities, and, subject to the above, to provide opportunities for scientific research to be undertaken in relatively undisturbed areas.
National Park Zone (green) Comparable IUCN category – II	To provide for the protection of the natural integrity and values of areas of the Marine Park, generally free from extractive activities, and, subject to the above, to provide opportunities for certain activities, including the presentation of the values of the Marine Park, to be undertaken in relatively undisturbed areas.
Preservation Zone (pink) Comparable IUCN category – IA	To provide for the preservation of the natural integrity and values of areas of the Marine Park, generally undisturbed by human activities.
Commonwealth Islands Zone	To provide for the conservation of the natural integrity and values areas of the Marine Park above low water mark, to provide for use of the zone by the Commonwealth, and, subject to the above, to provide for facilities and uses consistent with the values of the area.

Rezoning under the Representative Areas Program

The Representative Areas Program rezoned the entire Marine Park during a single comprehensive planning process that involved extensive consultation and lasted from 1999 to 2004. The primary aim of the programme was to better protect the range of biodiversity in the Great Barrier Reef by increasing the extent of the 'no-take' area, ensuring that they included 'representative' examples of all the different habitat types. A further aim was to minimize the impacts on the existing users of the marine park. In addition, growing

Table 4.2 *Activities matrix indicating which activities can occur in which zone, which are prohibited and which need a permit*

Permitted Activity	General use	Habitat Prot.	Cons. Park	Buffer	Scientific Res.	Mar. Nat Park	Preservation
Aquaculture	P	P	P	X	X	X	X
Bait netting	Y	Y	Y	X	X	X	X
Boating, diving, photography	Y	Y	Y	Y	Y	Y	X
Crabbing	Y	Y	Y	X	X	X	X
Harvest of aquarium fish, coral, beachworm	P	P	P	X	X	X	X
Harvest of sea cucmber, trochus, lobster	P	P	X	X	X	X	X
Limited collecting	Y	Y	Y	X	X	X	X
Limited spearfishing	Y	Y	Y	X	X	X	X
Line fishing	Y	Y	Y	X	X	X	X
Netting	Y	Y	X	X	X	X	X
Research	P	P	P	P	P	P	P
Shipping	Y	P	P	P	P	P	X
Tourism	P	P	P	P	P	P	X
Traditional use	Y	Y	Y	Y	Y	Y	X
Trawling	Y	X	X	X	X	X	X
Trolling	Y	Y	Y	X	X	X	X

Note: P denotes activities requiring a permit; Y denotes permitted activity; X denotes prohibited activity.

international awareness of the need for protection of biological diversity and representative examples of all habitats provided further impetus to the decision to rezone the entire GBRMP (Woodley, 2007).

These most recent rezoning plans have shifted the focus squarely on conservation, according to Jon Day (see Box 4.1). The rezoning was based on a recognized need to protect representative examples of each of the GBRMP's 70 bioregions. The zoning plan that was completed in 2004 is significantly different from the original zoning plans in all sections of the park. Whereas no-take or strictly protected areas accounted for only 4.6 per cent of the total GBRMP area in the previous zoning plans, the current plan has resulted in 33.3 per cent of the GBRMP being set aside as no-take or highly protected zones (e.g., in Marine National Park Zones commonly known as 'green' zones and small preservation zones or 'no-entry' areas). The official designation of these 117,000km^2 as off-limits to fishing and collecting was approved by the Australian Parliament in 2004. A further one-third is protected from activities that would impact the benthic habitat (including habitat protection zones, conservation park zones and buffer zones). The conservation focus extends to

Table 4.3 *Biophysical and socioeconomic-cultural operating principles used by GBRMPA*

Biophysical operational Principles

1. Have no-take areas the minimum size of which is 20km along the smallest dimension, except for coastal bioregions
2. Have larger (versus smaller) no-take areas
3. Have sufficient no-take areas to insure against negative impacts on some part of a bioregion
4. Where a reef is incorporated into no-take zones, the whole reef should be included
5. Represent a minimum amount of each reef bioregion in no-take areas
6. Represent a minimum amount of each non-reef bioregion in no-take areas
7. Represent cross-shelf and latitudinal diversity in the network of no-take areas
8. Represent a minimum amount of each community type and physical environment type in the overall network
9. Maximize use of environmental information to determine the configuration of no-take areas to form viable networks
10. Include biophysically special/unique places
11. Include consideration of sea and adjacent land uses in determining no-take areas

Social, economic, cultural and management feasibility operational principles

1. Maximize complementarity of no-take areas with human values, activities and opportunities
2. Ensure that final selection of no-take areas recognizes social costs and benefits
3. Maximize placement of no-take areas in locations which complement and include present and future management and tenure arrangements
4. Maximize public understanding and acceptance of no-take areas, and facilitate enforcement of no-take areas

all bioregions: under the plan, at least one-fifth of each bioregion is covered by multiple green zones with no extractive use.

A comprehensive programme of rezoning was finalized in 2004, after extensive and unprecedented public involvement. Public participation in the planning effort and via comments on the draft plan was intense, with some 31,000 submissions being considered during the rezoning process. According to GBRMPA, the rezoning process to provide for better 'representativeness' was strongly based on the best available science – biological, physical, social, economic and cultural.

The rezoning plan that was completed in 2004 is significantly different from the original plan in all sections of the park. In addition, growing international awareness of the need for protection of biological diversity and representative examples of all habitats provided further impetus to the decision to rezone the entire GBRMP (Woodley, 2007).

The rezoning process

The process of rezoning involved extensive use of science, analytical tools to process data and deal with public submissions and skilful management of the political and social concerns that rezoning raised. Prior to evaluating rezoning options, the GBRMPA embarked in a lengthy process to identify 70 distinct bioregions (30 reef bioregions and 40 non-reef bioregions). The rezoning was then guided by a set of biophysical operating principles and social-economic-cultural operating principles, shown in Table 4.3, that were used to allow the choice of final no-take zones to meet the aims of ensuring protection of all bioregions in no-take zones while minimizing the impact on reef users.

The new zoning plan increased the proportion of 'no-take' areas from less than 5 per cent to more than 33 per cent; however, the application of the 'complete package' of all operational principles listed in Table 4.3 ensured this no-take network was spread across all bioregions that; at least 20 per cent or more of every bioregion was in no-take areas; there was replication within bioregions; biophysically special and unique features were included; and crossshelf and latitudinal diversity were represented in the network (Osmond et al., 2010).

The outlining of bioregions, the selection of operating principles and the process of evaluating rezoning options were made very transparent by the GBRMPA, and the public was encouraged to submit their individual ideas on what areas might be important to protect. The process of public consultation set new standards for involvement in decision-making, and is now being used as a springboard to a more regionalized approach to community engagement as the new zoning plan is put into effect. Over 30,000 submissions were received in the rezoning process, resulting in a significantly longer planning process than the GBRMPA originally anticipated.

The most recent rezoning of the GBRMP was a complex, lengthy and resource intensive project that could probably not serve as a model for other initiatives of a similar, or smaller scale. And it should be noted that not all people were pleased with result and a review is being undertaken to seek improvements in decision-making processes. GBRMPA recognizes that widespread public support and compliance is essential for the successful implementation of the new plan. Nonetheless, the process is informative, especially as it highlights the power and utility of two new tools for zoning: MARXAN, the Reserve Design software chosen by GBRMPA to assist in placement of zones, and the system to evaluate resulting networks against core sociological and socioeconomic goals (Fernandes et al., 2005).

It is important to note that although MARXAN is a decision-support software, it will not tell the user the optimal solution. MARXAN yields two types of maps as a result of running algorithms that capture various sorts of zoning/management scenarios: best solution maps and 'irreplaceability' maps. By running each scenario many times, it is possible to decipher the percentage of runs in which each planning unit was picked, resulting in a value of irreplaceability; these maps show the relative importance of areas for meeting targets, and allow for prioritization between areas. The best solution maps, on the other hand, show solutions with the lowest total cost of implementing the zoning scheme (Vincent et al., 2004). Since public acceptance of zoning is so critical to the success of zoning initiatives, it is imperative that the planners get public feedback on proposed zoning plans and amend them as necessary to satisfy the greatest number of stakeholders (including decision-makers and management agencies that will be footing the bill) (Agardy, 1997).

Legal framework for zoning within the GBRMP

The comprehensive review of the Marine Park that was completed in 2006 has led to the reformulation of the Act that legally establishes the GBRMP and the GBRMPA. The amended Act states:

1 The main object of this Act is to provide for the long term protection and conservation of the environment, biodiversity and heritage values of the Great Barrier Reef Region.
2 The other objects of this Act are to do the following, so far as is consistent with the main object:
   a allow ecologically sustainable use of the Great Barrier Reef Region for purposes including the following:
      i public enjoyment and appreciation;
      ii public education about and understanding of the Region;
      iii recreational, economic and cultural activities;
      iv research in relation to the natural, social, economic and cultural systems and value of the Great Barrier Reef Region;
   b encourage engagement in the protection and management of the Great Barrier Reef Region by interested persons and groups, including Queensland and local governments, communities, Indigenous persons, business and industry;
   c assist in meeting Australia's international responsibilities in relation to the environment and protection of world heritage (especially Australia's responsibilities under the World Heritage Convention).
3 In order to achieve its objects, this Act:
   a provides for the establishment, control, care and development of the Great Barrier Reef Marine Park; and
   b establishes the Great Barrier Reef Marine Park Authority; and
   c provides for zoning plans and plans of management; and
   d regulates, including by a system of permissions, use of the Great Barrier Reef Marine Park in ways consistent with ecosystem-based management and the principles of ecologically sustainable use; and
   e facilitates partnership with traditional owners in management of marine resources; and
   f facilitates a collaborative approach to management of the Great Barrier Reef World Heritage area with the Queensland government.

Excerpted from the *Great Barrier Reef Marine Park Act 1975*, s. 2A

These legislated objectives were the genesis for the specific objectives of many of the zones and have been periodically updated to those that appear in the statutory zoning plan today (see Table 4.1). It is interesting to note that as the park itself matured and evolved, so too did its goals and objectives, and rezoning the entire protected area was the inevitable consequence.

However, it must be noted that while zoning is a key management instrument for the conservation and management over the entire GBRMP, there are many misconceptions about the role that zoning alone plays in the GBRMP (Day, 2008). Zoning does provide a spatial basis for determining where many activities can occur, but zoning is only one of many spatial management tools used in the GBR. Furthermore, zoning is not necessarily the most effective way to manage all ocean activities, and some are better managed using other spatial and temporal management tools (Day, 2008).

Some of the other management 'tools' or strategies applied in the GBR Marine Park include:

- Permits (often tied to specific zones or smaller areas within zones, and providing a detailed level of management arrangement not possible by zoning alone)
- Statutory 'Plans of Management' (legislative)
- Site Management Plans
- Special Management Areas (see below)
- Other spatial restrictions (e.g., Defence Training Areas, shipping areas, Agreements with Traditional Owners)
- Best Environmental Practices
- Industry Codes of Practice
- Partnerships with Industry (for example, the GBRMPA and the commercial fishing industry have recently commenced a partnership relating to 'future proofing' the industry in the face of climate change)

Many of these spatial management tools are superimposed on the underlying zoning, but with their own specific objectives (Day, 2009). Special Management Areas (SMAs), for instance, restrict use or access within specific areas of the Marine Park, providing a responsive and flexible management approach at a site-specific level. SMAs may be declared on a temporary, seasonal or permanent basis for a number of reasons including:

- conservation of a species or natural resource, e.g., turtles, dugong, bird nesting sites or fish spawning aggregation sites
- protection of cultural or heritage values
- public safety
- appreciation by the public
- as a response to emergency situations requiring immediate management action (e.g., a ship grounding, oil spill or marine pest outbreak).

The fact that the GBRMP utilizes a multilayered spatial management approach does mean the total management 'package' appears complex, but it does not require every aspect to be shown to the public in a two-dimensional zoning plan. Rather, the differing spatial layers need to be well understood by the managers and the specific user group for which the management measure has been developed. Similarly some spatial management tools need not be permanently in place to achieve effective results and do not appear as part of the zoning plan (e.g., the temporal closures that are legislated across the entire Great Barrier Reef prohibit all reef fishing during specific moon phases when reef fish are spawning).

Misconceptions

Misconceptions surrounding the planning of marine protected areas and the zoning within them can derail good planning and management processes. In the case of the Great Barrier Reef Marine Park, the investment in public awareness and education has been substantial. Nonetheless, mistaken impressions remain. Jon Day summarizes these misconceptions in a Perspective piece written for the newsletter *Marine Ecosystem and Management* (MEAM) (Day, 2008). His article appears in Box 4.1.

Box 4.1 Clarifying misconceptions about zoning – the GBR example

Jon Day, GBRMPA

Most people associated with managing marine or coastal areas have heard of the Great Barrier Reef (GBR); many also know that the GBR is an extremely large Marine Park and may even be aware that there are different zones that prohibit various activities in certain areas. There are, however, many misconceptions about the zoning scheme.

Zoning provides a spatial framework for managing use while providing differing levels of protection for different areas. In some parts of the world, zoning is based solely around allowing, or prohibiting, specific activities in specific areas. In the GBR the emphasis is on providing a spectrum of zones with differing objectives (the objectives and the provisions in the zoning plan then clarify what activities are appropriate in the zone and which of those activities will need a permit).

There are seven zones: the General Use Zone allows the widest range of marine activities, and the Preservation Zone the fewest. Under each objective for each zone, the GBR Zoning Plan sets out in detail two specific lists of 'use or entry' provisions; these then help to determine what activities may be allowed to occur in that particular zone. If an activity is not listed in either (1) or (2) above, then it is effectively prohibited. For communication with the wider public, the allowable activities in the Zoning Plan have been made into a simple activity/zoning matrix.Contrary to popular belief, not all activities listed in the Zoning Plan can occur with a permit. For example, activities such as aquaculture or harvest fishing that may or may not have an impact MAY be permitted even in Habitat Protection Zones, but only after undergoing a detailed permit assessment process.

There is also a special 'catch-all' permit provision in the GBR Zoning Plan ('any other purpose consistent with the objective of the zone') that provides for new technology or activities that were not known when the Zoning Plan was approved and which therefore are not in either of the above lists. In all zones an activity that is not specifically listed but is deemed to be consistent with the zone objective may be considered for a permit. This essential requirement (i.e. to be consistent with the zone objective) is an important first consideration whenever a permit application is considered for a particular zone. The 'any other purpose' provision is an important 'safety net' that has enabled new activities to be considered while maintaining the integrity of the overall zoning scheme.

Another misconception is that the GBRMP was zoned when implemented. The first zoning plans were progressively developed for parts of the Marine Park in the early 1980s, but it was not until 15 years after the Marine Park was declared (i.e., 1988) that the entire area was zoned. From 1988 until mid-2004, less than 5 per cent of the entire GBR was zoned in highly protected 'no-take' zones. The Representative Areas Program did rezone the entire Marine Park during a single comprehensive planning process, and today one-third of the GBR (i.e., 117,000km^2) is now highly protected. A further one-third is protected from activities that would impact the benthic habitat (including Habitat Protection Zones, Conservation Park Zones and Buffer Zones).

While zoning is a key management instrument for the conservation and management over the entire area (344,400km^2), another misconception is the role that zoning plays in the GBR Marine Park. Zoning does provide a spatial basis for determining where many activities can occur, but zoning is only one of many spatial management tools used in the GBR. Some of the other management 'tools' or strategies applied in the GBR Marine Park include:

- Permits
- Statutory 'Plans of Management'
- Site Plans/Special Management Areas
- Other spatial restrictions (eg Defence Training Areas, Shipping Areas, Agreements with Traditional Owners)
- Best Environmental Practices/Codes of Practice
- Partnerships with Industry

A final misconception is that zoning and hence management of the Marine Park is confined to just the 'wet bits' in the Marine Park. Legislative controls apply equally to the airspace above the Marine Park (up to 3000') as well as into the seabed; and in November 2004, the State of Queensland 'mirrored' the Commonwealth zoning in most of the adjoining State waters, so there is now complementary zoning for virtually all the State and Commonwealth waters within the entire GBR World Heritage Area.

Implications of zoning within MPAs

The GBRMP rezoning remains the best example of large-scale zoning found anywhere in the world. In part due to its size and complexity, the Australian example also shows how conservation values can be objectively balanced with development values (Agardy, 1997). The GBRMP has thus retained its multi-use character and its evolution over time, especially through the process of zoning and successive rezoning, makes management of the region ever more integrated (Osmond et al., 2010). For example, the Australian government has established shipping zones which guarantee access for transportation entities, irregardless of future conservation efforts (in a sense these can be thought of as sacrificial areas, which were selected in the kind of reverse selection process alluded to in Chapter 1).

The objectives-driven marine spatial planning process embodied in the GBRMP zoning and rezoning efforts has spurred other efforts elsewhere in

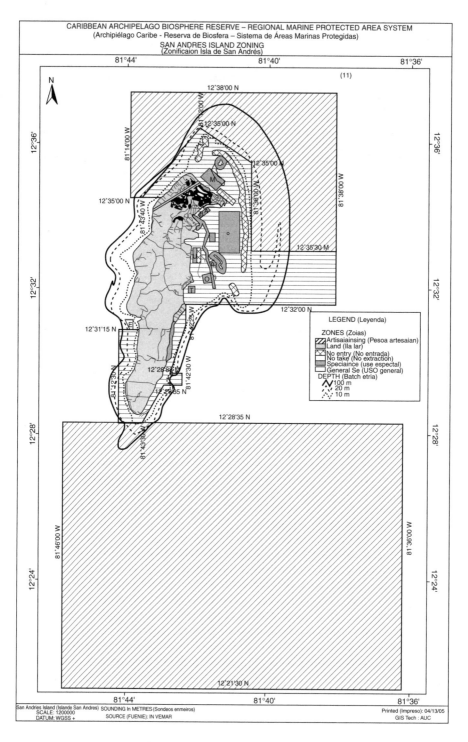

Figure 4.1 *San Andrés Zoning Plan within Seaflower Marine Protected Area*
Source: Coralina

Australia and has served as a model for zoning initiatives in other areas in the world. In the state of South Australia, a new large-scale, ecosystem-based zoning policy for management of development and use in the marine environment has been instituted with ocean zoning at its core. Planners are looking at four possible ecologically rated zones, with a series of goals, objectives and strategies representing the desired outcomes for each of the ecologically rated zones. The state is using a Performance Assessment System (PAS) to evaluate the success of the marine plans, matched to the specific objectives determined for each zone (Day et al., 2008).

Other very large-scale marine protected areas outside the continent of Australia also deserve mention for the innovative ways they are using ocean zoning to manage vast areas and a wide array of uses. Most notable is the Seaflower Marine Park in the San Andrés Archipelago of Colombia – the seventh largest marine protected area in the world and the largest in the Caribbean region. The Seaflower MPA, as well as the Biosphere Reserve within which the MPA sits, have a complex yet successful zoning plan, a portion of which is shown in Figure 4.1

The 65,000km^2 Seaflower MPA is divided into three administrative sections. The protected area, which grew out of vast stakeholder and community consultation and is administered by a Colombian governmental corporation known as Coralina, promotes an integrated, sustainable development approach to conservation. The objectives of the MPA are:

1 Preservation, recovery, and long-term maintenance of species, biodiversity, ecosystems, and other natural values including special habitats
2 Promotion of sound management practices to ensure long-term sustainable use of coastal and marine resources
3 Equitable distribution of economic and social benefits to enhance local development
4 Protection of rights pertaining to historical use
5 Education to promote stewardship and community involvement in management

To realize these objectives, the MPA has five types of zones:

1 Artisanal fishing (traditional methods and users only)
2 No entry (research and monitoring only)
3 No take (non-extractive activities only)
4 Special use (e.g., ports, shipping lanes, cruise-ship anchorage, intensive water sports areas, etc.)
5 General use. This zoning plan is similar to that of the Great Barrier Reef Marine Park, and the planners in Colombia have benefited from exchange of information with the GBRMPA, as well as with other MPA practitioners

Seaflower MPA zoning is relatively new, and amendments continue to be made. This case certainly deserves its own chapter in a book on ocean zoning, but because of its relative newness, lessons learned from the nacsent efforts are incorporated here. Coralina Director Dr Elizabeth Tayor and Coralina Conservation Advisor Dr Marion Howard are assessing the process and finalizing a paper on the zoning process, which should be published in 2010.

The most valuable principles that the GBRMPA and other zoning efforts within large-scale MPAs offer are twofold: 1) that use of marine spatial planning and ocean zoning should be tied to very specific, and clearly articulated, objectives; and 2) that zoning is a living, iterative process that must adapt to changing times, environmental conditions, and societal expectations.

Acknowledgment

The descriptions of the Great Barrier Reef planning and management processes in this chapter drew considerably on the input of Jon Day, one of the directors within the GBRMPA, who was responsible for commencing the Representative Areas Program that led to the rezoning. Jon has also assessed and analysed the progress that the GBRMP has made with respect to its original mandate and evolving conservation and societal objectives in a review for a forthcoming publication on MPA governance.

References

Agardy, T. S. (1997) *Marine Protected Areas and Ocean Conservation*, R. E. Landes Co, Austin, TX, USA

Day, J. C. (2002) 'Zoning – Lessons from the Great Barrier Reef Marine Park', *Ocean and Coastal Management*, vol 45, pp139–156

Day, J. C. (2008) 'EBM Perspective: Clarifying misconceptions about zoning – the GBR example', *MEAM*, vol 2, no 1, pp6

Day, J. (2009) 'Governance of the Great Barrier Reef Marine Park', paper prepared for a workshop on MPA governance, 11–16 October 2009, Lošinj, Croatia

Day, V., R. Paxinos, J. Emmett, A. Wright and M. Goeker (2008) 'The Marine Planning Framework for South Australia: A new ecosystem-based zoning policy for marine management', *Marine Policy*, pp535–543

Fernandes, L., J. Day, A. Lewis, S. Slegers, B. Kerrigan, D. Breen (2005) 'Establishing representative no-take areas in the Great Barrier Reef: large-scale implementation of theory on marine protected areas', *Conservation Biology*, vol 19, no 6, pp1733–1744

GBRMPA (2003) *Great Barrier Reef Zoning Plan 2003*, GBRMPA, Canberra, Australia

GBRMPA (2007) available at www.gbrmpa.gov.au/corp_site/management/representative_areas_program/rap_publications/info_sheets

GBRMPA (2009a) *Draft Corporate Plan 2009–2014*, GBRMPA, Canberra, Australia

GBRMPA (2009b) *Great Barrier Reef Outlook Report 2009: In Brief*, GBRMPA, Canberra, Australia

Kay, R. and J. Alder (2005, 2nd edition) *Coastal Planning and Management*, Taylor and Francis, Abingdon, UK, and New York

Osmond, M., S. Airame, M. Caldwell and J. Day (2010) 'Lessons for marine conservation planning: A comparison of three marine protected area planning processes', *Ocean and Coastal Management*, vol 53, no 2, pp41–51

Vincent, M. A., S. M. Atkins and C. M. Lumb (2004) 'Irish Sea Pilot: Report on the Communications Strategy', Joint Nature Conservation Committee, Peterborough, UK, available at www.jncc.gov.uk/irishseapilot

Woodley, S. (2007) 'Great Barrier Reef Marine Park Case Study', Study to support the Marine Protected Area Module, Center for Biodiversity Conservation, American Museum of Natural History, New York

5
Various Incarnations of Ocean Zoning in New Zealand

Tundi Agardy

Is the beach considered Crown land?

A checkered history

Tracking New Zealand's effort in ocean zoning has been challenging in that it was one of the first countries to declare a national commitment to ocean zoning of its entire territorial sea and Exclusive Economic Zone (EEZ), only to retreat to a much more circumspect use of zoning today. At the turn of the millennium, there seemed to have been a strong push for comprehensive national zoning, and planners appeared to have gone behind closed doors to develop a rather top-down marine spatial plan. However, the government met considerable resistance from the fishing industry, as well as from aboriginal groups. When the New Zealand government retooled and attempted much smaller scale zoning and protected area site selection, many environmental groups were outraged by the perception that conservation had slipped in priority – and the government seemed to find itself in a no-win situation. Recently, however, a more ground-up zoning initiative seems to have rekindled both government and public enthusiasm for ocean zoning.

Marine zoning is provided for in New Zealand's environmental management in the territorial sea (to 12 nautical miles [nm]), but according to Wallace and Weeber (2008), it is primarily used to control and to provide for aquaculture. Marine protected areas have been instituted throughout the country's nearshore, but these have been planned and implemented in a rather ad hoc manner, and their siting neither corresponds to articulated goals and objectives, nor to a developed national strategy. Further offshore, New Zealand's environmental management is patchy and many sectors, such as offshore mining, lack adequate environmental controls, at least according to conservation interests.

At the time of writing, New Zealand legislators were wrapping up several years of work on the new Exclusive Economic Zone Environmental Effects (EEZEE) Bill that aims to safeguard the integrity of the country's ecosystems by 'bringing in new mechanisms to monitor and manage marine activities' (Ministry for the Environment [MFE], 2009). This new legislation may well incorporate marine spatial planning to achieve more holistic, ecosystem-based management, but it is too early to tell if the EEZEE will utilize ocean zoning to its fullest potential.

Spatial management in New Zealand's EEZ

New Zealand's Exclusive Economic Zone is one of the largest EEZs in the world, with a wealth of natural biodiversity values, and, according to New Zealand's Ministry for the Environment, great potential for future economic opportunities such as seabed minerals and energy generation (MFE, 2009). The government of New Zealand has publicly stated that the lack of a comprehensive regulatory system constrains the country from harnessing this potential while managing the environment in a sustainable way.

This recognition of the need for a comprehensive framework led to a series of high level government initiatives, including some false starts in policy development. As early as 2002, when the Oceans Policy Secretariat commissioned an Oceans Issues paper to investigate ocean use rights, the inadequacy of marine management in New Zealand was noted and calls for a new mechanism to resolve ocean use conflicts emerged (MFE, 2003a).

The New Zealand government then embarked on a three-stage process for addressing ocean use rights. At the time, there was much publicity about using comprehensive ocean zoning to strategically and efficiently allocate uses and minimize conflicts. These stages were to have been:

- Stage One – Designing the vision
- Stage Two – Developing options to achieve the vision (We are still working through this stage. See our current oceans work for details.)
- Stage Three – Implementing the vision

The process only got to Stage Two before it was aborted, thanks to a change in focus of Oceans Policy development.

Two reports released in 2005 kickstarted a second process for developing a coherent policy framework for New Zealand's EEZ. These are:

- *Getting Our Priorities Right* – exploring the role of information in setting national priorities under an Oceans Policy
- *Offshore Options: Managing Environmental Effects in New Zealand's Exclusive Economic Zone* – examining gaps in environmental controls in the Exclusive Economic Zone (the area between 12 and 200nm offshore)

In August 2007 the Ministry for the Environment released a discussion paper, guided by the draft oceans policy framework that the ministry had developed. The discussion paper sought comment on a preferred legislative option for managing the impacts of activities in the EEZ. The Ministry for the Environment then developed more detailed policy for improving regulation of marine activities in the EEZ in late 2007/early 2008 (MFE, 2009). In June 2008, drafting began on the Exclusive Economic Zone Environmental Effects Bill; unfortunately, the drafting of the Bill was not completed before the 2008 general election, and the government thus experienced another delay. According to the Ministry for the Environment website, completing the drafting is a priority for the ministry (at the time of writing the drafting was still not complete, though it was expected to be finished and presented before the Select Committee before the end of 2009).

As mentioned above, law under development for the EEZ may well include zoning to achieve ecosystem based management goals, but the policy formulation process has not been very transparent and it is difficult to predict the outcome.

Spatial management within New Zealand's territorial sea

Management of coastal lands and the marine environment within 12nm is under the purview of New Zealand's Resource Management Act of 1991

(RMA). According to Cath Wallace (2008), the RMA embraces the principle of sustainability and provides for the sustainable management of natural and physical resources, constituting a major reform of New Zealand's terrestrial and coastal environmental administration. The Act and the subsidiary New Zealand Coastal Policy Statement stipulate that an effects-based approach with open public participation be utilized in coastal management. In principle, consistent processes covering human activities that affect air, land and water achieve integrated management through a process of national, regional and district policies and plans. Zoning is one tool to achieve the integrated management, but it is not mandatory. However, in 2005 an amendment directed agencies to evaluate certain areas for appropriateness of aquaculture development, with the resulting establishment of Aquaculture Management Areas. According to Wallace and Weeber (2008), this special use of ocean zoning grew in response to a 'gold rush' of marine farming applications, which threatened to overwhelm the country's coastal management councils as well as the coastal environment. To date, this is the only use of zoning in territorial waters, with the exception of marine reserves, and fisheries closures.

Fisheries and ocean zoning in New Zealand

Fisheries management in New Zealand is known the world over due to the country's early and steadfast commitment to tradable quota systems, known as Individual Transferable Quotas (ITQs) or catch shares.

At the time that the Resource Management Act was beginning to be developed in the middle to late 1980s, it was decided that fisheries management would not be included because New Zealand had already embarked on the property rights experiment that became the Quota Management System (QMS). Instead, most environmental impacts of fishing are regulated under the Fisheries Act of 1996, which establishes Quota Management Areas, quota limits such as Total Allowable Catch (TAC) and Total Allowable Commercial Catch (TACC). Sustainability measures include method, gear, season and area restriction, and fishing boat size limits in some areas, and the Ministry of Fisheries is developing species-based fisheries plans for selected species.

According to Wallace and Weeber (2008), proposals for marine reserves, spatial management and controls on fishing generally encounter vigorous opposition. They contend that property rights, held in ITQs, empower opposition to conservation measures and have made the politics of spatial management and protection much more difficult. However, a recent op-ed in the *New Zealand Herald* by Owen Symmans, formerly the Chief Executive Officer for the New Zealand Seafood Industry Council, suggests that fishing interests could well rally behind comprehensive ocean zoning to stave off perceived threats to access and use of ocean space and resources by other sectors. The op-ed is reprinted in its entirety in Box 5.1 (Symmans, 2009).

Such a view, together with the fact that fisheries quota areas could well constitute a foundation for using ocean zoning to help increase the efficacy of fisheries management, suggests that a meeting of minds between fisheries interests, conservationists and regulators could be within reach.

Box 5.1 Foresight needed to manage ocean treasures

Owen Symmans, former chief executive, New Zealand Seafood Industry Council

Take the area of New Zealand, and multiply it by four. Add spectacular mountain ranges, yawning canyons, erupting volcanoes and some of the world's most pristine, untouched and unexplored wilderness. Now imagine that this area has been recognised as part of New Zealand's natural heritage and has been protected from commercial exploitation by law, forever.

If that sounds like the kind of thing a bunch of conservationists might dream up, that's because it is.

Two years ago, 1.2 million sq km of seabed around New Zealand was set aside as regulated benthic protection areas, or BPAs. These areas – a total of 17 different ones within the country's exclusive economic zone – are protected from all bottom-trawl fishing methods, including dredging. The areas are broadly representative of marine biodiversity in New Zealand waters.

It's hard to appreciate how significant this action was, but consider this – it increased the area of New Zealand's protected seabed from 3 per cent of our exclusive economic zone to 32 per cent overnight. These areas are a national treasure, something to be proud of. But not many people know they even exist, and I'd venture that a lot fewer would know who was behind them – New Zealand's commercial fishing industry.

Why would the group of people with so much to gain from fishing our deep water work with the Government to ensure it never happens in these places? Well, it makes more sense than you think.

The answer goes to the heart of the way we look at our business. More than any other major natural resource-based industry in New Zealand, our economic destiny is inextricably linked to our environmental practices. To put it bluntly, if we don't look after our resources, we won't have an industry.

We pushed for protection areas because we wanted to find the balance that would allow New Zealand's $1.6 billion a year commercial fishing industry to continue to operate and export, while at the same time responding to concerns about protecting the marine environment. It was the right thing to do.

That's why when I read Anthony Doesburg's comment (Hot prospects in deep water) I felt it needed a response. Mr Doesburg urges Minister Gerry Brownlee to look into the possibility of deep water mining, before 'we start ripping up national parks in search of mineral wealth'.

On one level that's an understandable position. After all, mining under the sea takes place out of sight, and would undoubtedly cause less public controversy than mining in a national park.

The irony for the seafood industry is that the very area Mr Doesburg is proposing – along the Kermadec arc, which stretches northeast of New Zealand between the Bay of Plenty and Tonga – is within a benthic protected area.

That area, called the Tectonic Reach area, includes a high concentration of seamounts and hydrothermal vents, which allow the minerals to be deposited.

Industry and government, backed by science, agree that we can still harvest fish sustainably from such areas using fishing methods that do not go near the bottom, and we continue to do so.

Bottom trawl and dredge fishing remain essential fishing methods for a large proportion of our commercial fisheries in areas outside of BPAs. It should be noted that less than 10 per cent of New Zealand's exclusive economic zone has ever been bottom-trawled.

> It is clear that New Zealand needs the income that comes from both fishing and mining. It's also clear that this needs to be achieved with due consideration of the effects we are having on the environment.
>
> That fact that the Kermadec arc can be proposed as a mining area in apparent good faith highlights a wider issue. We need a policy framework that governs all the different competing interests in our marine environment – seafood, mining, oil exploration, telephone and internet cable laying or whatever else might come along.
>
> Such an 'oceans policy' would ensure all industries working in the marine environment are able to achieve the right balance between their competing commercial interests and the need to safeguard environmental values.
>
> New Zealand's fisheries quota management system offers a model for this kind of balance. Just this year, independent research published in the journal Science singled New Zealand out for praise, because we acted with foresight to protect our fisheries before drastic measures were needed.
>
> With the fantastic commercial potential for new industries operating in the marine environment, we need to find a way to apply that foresight together.

Forward momentum: New Zealand in the context of international marine protection

The Ministry for the Environment's first background paper created to support the developing marine policy was entitled 'Setting the Stage: New Zealand's Oceans-Related Obligation and Work on the International Stage' (MFE, 2003b). Topics covered in the background paper include:

- How international law is applied in New Zealand
- Jurisdictions
- New Zealand's legally binding obligations under international instruments
- New Zealand's non-legally binding obligations under international instruments
- Legislation that codifies international obligations
- Strategies that help implement international obligations
- Current activities under way in key international institutions with oceans responsibilities and functions
- United Nations Informal Consultative Process on Oceans and the Law of the Sea (UNICPOLOS)
- International Seabed Authority (ISA)
- International Tribunal for the Law of the Sea
- Commission on the Limits of the Continental Shelf
- The Food and Agriculture Organization (FAO)
- The International Maritime Organization (IMO)
- United Nations Environment Programme: Activities in Marine and Coastal Areas
- The Commission on Sustainable Development (CSD)
- World Trade Organization & Committee on Trade and Environment
- Regional institutions with oceans mandates

Thus New Zealand was not only aware of and abiding by the evolving body of international law that applies to marine resource use and ocean protection, it was indeed beginning to play a leadership role. The last decade saw an enormous amount of activity centred on marine spatial planning, primarily in New Zealand government agencies. However, the public sector seemed unprepared for the public backlash that followed.

Retreat in the face of conflict

The backlash against comprehensive management of New Zealand's EEZ and territorial sea has come from two main fronts: fishers and aboriginal peoples, each of whom see themselves as being entitled to special rights.

As mentioned above, the fishing community had shown significant opposition to the establishment of marine protected areas, and the addition of new marine protected areas appears to be blocked until 2013 due to fishing industry pressure (Wallace and Weeber, 2008). However, a bigger challenge for the New Zealand government seems to be engagement of the Maori communities in the marine planning process.

A recent article in the *New Zealand Herald* (Nippert, 2009) describes the situation in Opotiki, a small settlement on the coast tucked among sandy beaches and mangrove forests. In 2005 Whakatohea, the local *iwi* (tribe), made the first claim under the contentious Foreshore and Seabed Act for customary rights to 47 km of coastline.

The Whakatohea are seeking *kaimoana* and *kingitanga* rights – in essence, the right to collect seafood and manage the water environment. Some people outside the tribe fear that the right of access to the shore will be taken away if the Whakatohea claim is recognized, saying that the broad customary rights could be interpreted as conferring title. The case has been stuck in limbo after the Maori Party struck a governance deal with the National Party that required a review of the Act. According to the *Herald* article, Prime Minister John Key has said the law will almost certainly be repealed, but not until what replaces it has been worked out (Nippert, 2009). In the meantime, the Whakatohea could negotiate directly with the Crown for a settlement that includes putting the foreshore and seabed on the table, as occurred in a previous case brought by the Ngati Porou iwi on the east coast.

The article also mentions a parallel development off the coast slowly taking shape, which similarly changes the landscape for asserting or allocating rights to resources and space. Plans are in place to develop New Zealand's largest mussel farm and test lines are soon to be put into place. The promise of generating millions in income for the region, and employing 500 people in a district that suffers chronic unemployment, has both the local council and the iwi excited and hopeful. Yet access to the mussel farm waters will of course be denied to locals – be they Maori or Pakeha, even including those who currently have customary use.

According to the *New Zealand Herald* article, the aquaculture industry has sailed along unimpeded by the ructions of the legal status over the seabed and foreshore. Indeed, a 2004 decision has flagged 20 per cent of all marine farming space established since 1991 to be given to iwi authorities (Nippert, 2009). The government was reviewing submissions to a government technical advisory group on aquaculture policy at the time of writing.

In referring to the debate over foreshore access (exemplified by the Opotiki situation, the *New Zealand Herald* article states, 'Interestingly, in perhaps a signal of a change, the right of public access and the acknowledgement of customary rights have become 'guiding principles' rather than 'bottom lines' in this drafting process' (Nippert, 2009). It will be interesting to see how these guiding principles influence the final draft of the Exclusive Economic Zone Environmental Effects Bill and any other related or subsequent marine policies that emerge from government. After fully charging ahead in comprehensive ocean zoning, then beating a hasty retreat, where the government and New Zealand stakeholders finally land is anyone's guess.

References

Ministry for the Environment, New Zealand (MFE) (2003a) 'Ocean Use Rights', Draft Working Paper 2–14 March 2003, available at www.mfe.govt.nz/publications/oceans/stage-two-papers-feb03/02-ocean-use-rights.pdf

Ministry for the Environment, New Zealand (MFE) (2003b) www.mfe.govt.nz/publications/oceans/stage-two-papers-feb03/11A-background-information.pdf

Ministry for the Environment, New Zealand (MFE) (2009) Ministry for the Environment website www.mfe.govt.nz; last accessed December 31 2009

Nippert, M. (2009) 'The rights and wrongs over new access to the foreshores', *New Zealand Herald*, 8 November, available at www.nzherald.co.nz/nz/news/article.cfm?c_id=1&objectid106076961&pnum=1

Symmans, O. (2009) 'Foresight needed to manage ocean treasures', *New Zealand Herald*, 26 November, available at www.nzherald.co.nz/news/print.cfm?objectid=10611634

Wallace, C. (2008) Personal communication

Wallace, C. and B. Weeber (2008) 'The status of spatial management in New Zealand', *Marine Ecosystems and Management*, vol 2, no 1, p8

6
Zoning Efforts in United Kingdom Waters

Tundi Agardy

The Isles of Scilly, off southwestern England

Spatial planning at different scales

The waters of the United Kingdom have seen much marine spatial planning activity in the first decade of the new millennium. Now the recent passing of various bits of enabling legislation paves the way for the UK to move from somewhat vague marine planing to actual ocean zoning. Authorities are currently working up maps with attendant regulations to better manage a spate of activities in UK maritime waters, as well as on the coast.

Marine spatial planning, protected area site selection and ocean zoning initiatives have occurred under a variety of government agencies, spurred on by non-governmental organizations and civil society. However, even though these initiatives are all related, they have not all been coordinated. The situation in the UK is complicated, given the different authorities in England, Scotland, Wales and Northern Ireland, as well as the fact that the UK is a part of the European Union and therefore also performs planning according to EU Directives. In addition, different zoning initiatives have taken place in internal waters, within the 200 nautical mile fishery zone, in the area extending to the outer limit of the continental shelf (the UK analogue to the Exclusive Economic Zone), and in waters shared by the UK and its neighbouring countries (e.g., the Irish Sea between the UK and Ireland and surrounding the Isle of Man) (see Figure 6.1).

In debating the legislation that allows better and more coordinated management of marine resources, coastal access and ocean space, the parliament considered how best to devolve management authority while still keeping planning as strategic and holistic as possible. For the purposes of the Marine and Coastal Access Act of 2009 (for details see section below), the government decided that the internal waters of Scotland, Wales and Northern Ireland would be under local management authority, while the waters of England and the offshore waters of the UK as a whole (to the limits of the Fishery Conservation Zone or the continental shelf, whichever is further) would be under a new UK management authority known as the Marine Management Organisation (MMO). Figure 6.2 illustrates the breakdown of management authority being proposed under the Marine and Coastal Access Act.

Before providing further details on the enabling legislation for ocean zoning, an earlier initiative to develop zoning plans for one region of the UK will be discussed: the Irish Sea Pilot.

Irish Sea pilot and conservation zones

Agencies within the UK and its neighbouring countries have long recognized the utility in ocean zoning, though neither the terms ocean zoning nor marine spatial planning were commonly used prior to the turn of the millennium, when these terms appeared in the management literature and the European lexicon.

In 2002, the UK government decided to embark on a regional planning effort to identify both areas of conservation interest and areas with development

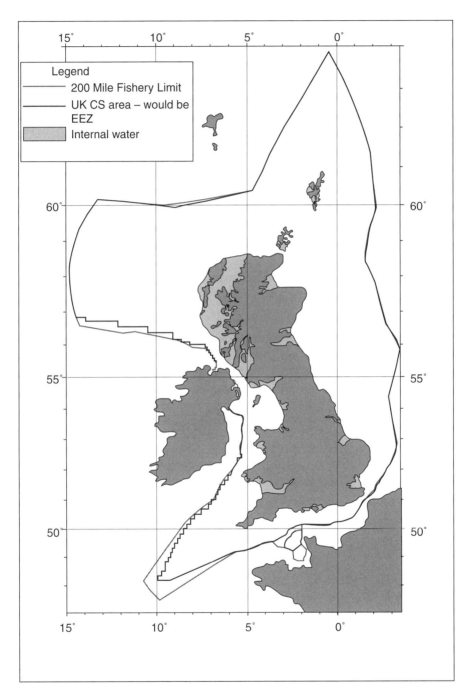

Figure 6.1 *Limits of United Kingdom internal waters, Fishery Conservation Zone and continental shelf*

Source: DEFRA, 2009

Figure 6.2 *Divison of various management authorities under the Marine and Coastal Access Act*

Source: DEFRA, 2009

potential for maritime industries. DEFRA, the UK Department of Environment, Food and Rural Affairs, and the Joint Nature Conservation Committee (JNCC, now called Natural England) signed a Service Level Agreement for a Regional Seas Pilot Scheme to be carried out in the Irish Sea. Working with the Countryside Council for Wales, the Environment and Heritage Service and the Government of the Isle of Man, DEFRA and the JNCC undertook data collection, assembled a database and analysed various planning options for the whole of the Irish Sea – publishing its findings in a 2004 report (Vincent et al., 2004).

Table 6.1 *Framework for marine nature conservation as elaborated in the Irish Sea Pilot*

Framework Elements	Considerations
Wider Sea	Issues of pollution, water quality, wide-ranging marine species and environmental change in all seabed and overlying waters under UK jurisdiciton
Regional Sea (e.g., Irish Sea)	Descriptions of biodiversity, identification of conservation priorities and management of human activities within ecologically meaningful subdivisions of the Wider Sea
Marine Landscapes (estuaries, rias, saline lagoons, sea lochs, sounds, gas structures, photic reefs, aphotic reefs, sea mounds, sand/gravel banks, coastal sediment deposits, shallow water mud basins, fine sediment plains, sediment wave/megaripple fields, low bed-stress coarse sediment plains, high bed-stress coarse sediment plains and deepwater channels)	Assessment of sensitivity of various habitat types, mapped using geophysical, hydrographical and ecological data
Habitats and Species	Identification of nationally important and/or sensitive areas to be the subject of special measures, including the identification of areas within which potentially damaging human activities would be strictly regulated

Source: DEFRA, 2004

DEFRA and JNCC used the Irish Sea Pilot as a test case for planning across several geographical scales, using a hierarchical framework that spanned the wider sea (UK waters), the regional sea (Irish Sea), marine landscapes and, finally, nationally important habitats and species (see Table 6.1). This nested framework was arguably the first time the British authorities had looked holistically at internal and external waters, for the purpose of identifying areas of critical importance for the future sustainability of marine resource and ocean space use. The Pilot collected and collated geophysical, hydrographical, nature conservation, ecological and human-use data and used GIS analysis along with decision-support software (MARXAN) to evaluate various options for spatial management.

The Irish Sea Pilot developed a base map for the coastline, as well as the 3, 6, and 12nm limits for the UK, Ireland and the Isle of Man, at scales spanning from 1:10,000 to 1:2,500,000. Additional data layers included hydrographical data (water temperatures, salinity, currents and frontal systems); bathymetry and seabed data; data on major vertebrates (seabirds, cetaceans, basking sharks and commercially exploited fish); data on benthic communities; data on distribution and relative intensity of fishing effort; data on ports, shipping routes and shipping intensities; data on oil and gas fields, locations of wells and pipelines, and surface structures; data on renewable energy development; data on sand and gravel extraction; data on coastal land use, including location and size of coastal settlements, developments with direct linkage to the Irish Sea, and Food and Environment Act consents; data on submarine cables; data on coast defence and flood defence structures; data on tourism and recreation; data on wastewater and industrial discharges; data on military practice and exercise areas; data on spoil disposal sites; and data showing locations of statutorily protected nature conservation sites.

Nationally important marine conservation areas were also identified, with the objective of eventually developing an ecologically coherent network of protected area sites having representative examples of each habitat type, areas of exceptional biodiversity and important areas for aggregrations of highly mobile species. Criteria for identifying such sites included typicalness, naturalness, size, biological diversity, critical areas for certain stages in the life cycles of key species and areas essential for a nationally important marine feature.

The Irish Sea is one of the UK's smaller regional seas, spanning 58,000km^2 in area. The primary contributors to the national and regional economies of the region are tourism and recreation, oil and gas, shipping and naval defence. According to the Pilot, the renewable energy sector is small but growing, while the fishing industry contributes only modestly to the region's economy. Using existing and potential development pressures and the data on biodiversity and ecological values, the Pilot was able to evaluate a number of different planning options. Using MARXAN to identify least-cost solutions, the Irish Sea Pilot yielded a series of maps, among them maps showing 'irreplaceability' (showing relative importance of areas for meeting conservation targets) and 'best solution' (lowest total cost) options for identifying nationally important sites (see Colour Plate 6.1). In theory, planners could then take this information and develop an ocean zoning map to allow appropriate uses in various areas within the region.

One of the main findings of the Irish Sea Pilot was that better coordination was needed – both in terms of data collection/management and in terms of identifying areas critical for conservation or sustainable development. It is likely that this test case for ocean zoning provided much of the foundation for the development of enabling legislation such as the Marine and Coastal Access Bill, and in particular the formation of the Marine Management Organisation. But the Irish Sea Pilot also led to a key regional project to identify priority areas for conservation – the Irish Sea Conservation Zones (ISCZ) project – as a first step towards developing a regional network of protected areas. This, too, could inform a regional zoning map at some point in the future.

The ISCZ project balances wildlife protection (including protection of the over 30 species of sharks that frequent the area) with commercial fishing, shipping, oil and gas, aggregates extraction and wind farm interests, along with recreation. The ISCZ is part of the UK government commitment to establish a network of marine protected areas by 2012 – as stipulated by the 2008 Marine Bill. Unlike other planning processes that precede it, the ISCZ is a stakeholder-led process, in which recommendations for protected areas will emerge from the users of that particular marine environment. The suggestions for siting marine conservation zones will need to meet scientific criteria, minimize social and economic costs and maximize benefits to society.

Such networks of marine protected areas in England can be considered an excellent starting point for comprehensive ocean zoning (see also the section on MPAs as a starting point in Chapter 3), since they can be comprised of five different types of sites: 1) Marine Conservation Zones, 2) Sites of Special Scientific Interest, 3) Special Areas of Conservation, 4) Special Protected Areas and 5) Ramsar Sites (wetlands of international importance). Whether the UK government will move from networks of protected areas to comprehensive ocean zoning remains to be seen, yet all the elements to do so appear to exist, including not only within such regional projects set up by Natural England and the Joint Nature Conservation Committee, but also the Marine Bill of 2008 and the recent Marine and Coastal Access Bill (now an Act).

The marine planning done under the Marine and Coastal Access Act and the earlier Marine Bill will encompass all marine activities and will focus not only on the different sectoral interests but also on the way activities interact, the conflicts between them and cumulative impacts.

The Marine and Coastal Access Act

The Marine and Coastal Access Act will lead to the establishment of a series of marine conservation zones around England and Wales and includes plans to create a footpath along the entire coastline of the two nations. Having been passed by the House of Lords, the Marine and Coastal Access Bill went to Buckingham Palace to receive Royal Assent, becoming an Act. In spring of 2009, Environment Secretary Hilary Benn indicated that the first stretch of coastal path and the first marine conservation zone would be established in 2012 – a target that was in fact reached prematurely with the 2010 announcement of England's first marine conservation zone (see below).

The intent of the Act is to conserve marine wildlife and steer a path towards sustainable uses of the seas (BBC News, 2009). In addition, the Act aims to minimize conflicts arising from competing pressures on ocean areas and resources. As well as the establishment of an uninterrupted coastal path and conservation zones, the Act also creates a government body called the Marine Management Organization (MMO).

According to DEFRA, the Marine and Coast Access Act intends:

> *to make provision in relation to marine functions and activities; to make provision about migratory and freshwater fish; to make provision for and in connection with the establishment of an English coastal walking route and of rights of access to land near the English coast; to enable the making of Assembly Measures in relation to Welsh coastal routes for recreational journeys and rights of access to land near the Welsh coast; to make further provision in relation to Natural England and the Countryside*

Council for Wales; to make provision in relation to works which are detrimental to navigation; to amend the Harbours Act 1964; and for connected purposes. (DEFRA website www.defra.gov.uk)

The Act stipulates that the UK government set up the new Marine Management Organisation as a centre of marine expertise, in order to provide a consistent and unified approach to marine planning, deliver improved coordination of information and data, and reduce administrative burdens (DEFRA, 2009). The Lords debated the merits of the prospective MMO in promoting efficiency and economies of scale – something that the status quo unintegrated management in the UK could not provide. The Marine Management Organisation will have functions spanning the entire management chain from planning through to enforcement.

DEFRA suggests that the Act will 'create a strategic marine planning system that clarifies the UK government's marine objectives and priorities for the future, and directs decision makers and users towards more efficient, sustainable use and protection of marine resources'. For instance, the changes that the Act will make to the marine licensing system will result in better, more consistent licensing decisions delivered more efficiently by a system that is proportionate and easier to understand and to use, integrating delivery across a range of sectors (DEFRA, 2009). DEFRA also states that the fisheries and environmental management arrangements are strengthened by the Act so that more effective action can be taken to conserve marine ecosystems and help achieve a sustainable and profitable fisheries sector. As part of modernizing inshore fisheries management in England, Sea Fisheries Committees (SFCs) will be replaced by Inshore Fisheries and Conservation Authorities (IFCAs) (DEFRA, 2009).

The first stage of this marine planning system will be the creation of a marine policy statement to create a more integrated approach to marine management and setting both short and longer-term objectives for sustainable use of the marine environment. The second stage will be the creation of a series of marine plans, which will implement the policy statement in specific areas, using information about spatial uses and needs in those areas (DEFRA, 2009). Movement on marine spatial planning has already begun. In a new development, England announced the creation of its first marine conservation zone on 12 January 2010. The identification of this site, around Lundy Island off the north Devon coast, marks the first in a series of protected zones that the four regional planning bodies established under the Marine and Coastal Access Act are expected to propose throughout UK waters.

Following the recommendations arising out of the Irish Sea Pilot, the Act highlights the importance of high quality marine data and the need for a sound evidence base for making informed policy and management decisions. The MMO will provide a renewed focus and centre of expertise for the collection, storage and accessibility of up-to-date data and information relating to the marine area. If comprehensive ocean zoning is a route that the government of the UK, or any of its more localized authorities, decide to take, the MMO will be perfectly poised to carry out the planning and implementation of comprehensive ocean zoning.

Scottish marine planning

The Scottish government recently considered proposals for new marine legislation in Scotland. It is expected that this will lead to the creation of a Scottish Marine Act, which would introduce new provisions for marine spatial planning and marine nature conservation.

A paper commissioned by Scottish Natural Heritage (2009) provides guidance on identification of particularly important areas to protect with a spatial management scheme. According to this document:

> *a suitably designed and managed marine planning system could make a major contribution to delivering natural heritage objectives. The hierarchical nature of spatial planning and the objectives-based approach provides an opportunity to embed ecosystem objectives and targets throughout the various levels of the plan. This will help to ensure that objectives are relevant at a local level and provide for engagement of local stakeholders. The application of zoning within marine plans provides an opportunity to protect habitats and species of conservation importance by ensuring that potentially damaging developments and activities are steered away from sensitive feature.*

Scottish stakeholders in government and civic society have assessed what components of a system of marine spatial planning are necessary to deliver nature conservation, including setting ambitious but feasible objectives and targets, policies and zoning. The paper commissioned by Scottish Natural Heritage suggests that it may be more appropriate to consider new provisions for marine nature conservation within an overall marine planning framework. The paper further stipulates what steps should be taken to support achievement of natural heritage objectives (see Box 6.1).

The Scottish Natural Heritage document also counsels on adapting a particular zoning mechanism without understanding the consequences, saying that the zoning can either be highly flexible or extremely rigid.

Identifying sites for a UK-wide marine reserve network

The efforts by the Scottish Parliament to develop a national marine Bill complement the efforts of the UK Parliament. Similarly, the advocacy by Scottish environmental groups is complemented by the work of the Marine Conservation Society (UK) in identifying marine areas of particular importance for wildlife. The location of 73 sites deemed important by the conservation group could inform a future zoning process by indicating possible sites for strictly protected zones within an ocean zoning scheme. At the time of writing,

> **BOX 6.1 RECOMMENDATIONS REGARDING SCOTTISH MARINE PLANNING RELATING TO ZONING, AS PROVIDED BY SCOTTISH NATURAL HERITAGE**
>
> The planning process should provide for a flexible and proportionate approach to marine nature conservation comprising:
>
> - Objectives and targets at national/regional and, where appropriate, local level
> - National/regional policies, supported by more local policies where appropriate
> - A system of zoning to protect nationally important marine areas (NIMAs) as appropriate, by zoning away human activities and developments which are incompatible with the natural heritage interest
> - Scope for formal nature conservation designation for 'best of' sites or in circumstances where specific and targeted management actions might best be delivered through this approach. These provisions would need to be clearly circumscribed to ensure that they could not be used to undermine the wider marine planning system
>
> The marine planning system should include specific provision for zoning that would govern the application of policies in defined areas, for example:
>
> - Defining types and levels of activity and development that would/would not be compatible with specific nature conservation requirements
> - Establishing policies within marine plans to steer potentially damaging developments and activities away from particularly sensitive features
> - Providing for the active monitoring and management of activities where capacity limits have been established
> - Providing for clear enforcement of zoning policies

there was only one small strictly protected bit of coastline in Britain – the sea around Lundy Island in the Bristol Channel, off North Devon.

According to the *Telegraph*, the recommendations made by the Marine Conservation Society would increase the amount of UK seas protected in marine reserves from just $3.7m^2$ (6km) to $132m^2$ (212km). Such protection appears to be supported by a large proportion of the British populace: a survey of 527,000 Co-op customers across the country found over 80 per cent of participants supported the introduction of marine reserves (*Daily Telegraph*, 2009).

Integration across land, sea and watersheds

In addition to these wholly marine-focused zoning initiatives, the UK has positioned itself to create truly integrated zoning plans that span land and sea. In a 2008 DEFRA publication, the remit of the government under the auspices of integrated management extends across a wide swath of coastal land and sea. UK Planning Policy Guidance Note 20 on Coastal Planning offers the following guidance for Local Planning Authorities in defining the coastal zone for their areas:

> It could include areas affected by off-shore and near-shore natural processes, such as areas of potential tidal flooding and erosion; enclosed tidal waters, such as estuaries and surrounding areas of land; and areas which are directly visible from the coast. The inland limit of the zone will depend on the extent of direct maritime influences and coast-related activities. In some places, the coastal zone may be relatively narrow, such as where there are cliffs. Elsewhere, particularly where there are substantial areas of low-lying land and inter-tidal areas, it will be much wider.

This need for integration is echoed by the European Commission (2007), which declared that 'working at different scales and across administrative and sectoral boundaries remains a formidable challenge, but is central to achieving integration'.

Marine plans developed at various scales but within the national policy framework can achieve this needed integration. Such marine plans can guide decisions on licence applications and other issues, and give coastal regulators and communities a way to input into marine planning in the way they already input into land planning at the coast. According to this DEFRA publication (2008), the benefits of effective marine planning include:

- bringing together coastal managers, users and communities and enable them to work together and shape the direction of plans from an early stage. This will allow coastal managers and land planners to feed their skills, experience and knowledge into marine plans
- people being offered a chance to be involved in the planning process, which will give them a greater sense of ownership of the final plan, and will increase the chance of the plan being implemented through effective decision-making
- a participatory planning approach from an early stage, laid down within the Statement of Public Participation, will ensure greater transparency and enable increased public understanding of the marine environment;
- the opportunity to highlight potential problems and conflicts between marine uses at an early stage in the development of marine plans, providing greater opportunity to resolve conflicts and prevent unnecessary delays
- the gathering and sharing of consistent information during the planning process to encourage marine and coastal regulators to make more consistent decisions
- the clear establishment of marine plan authorities in different parts of the UK's waters, and mechanisms to enable them to delegate those responsibilities

Integration will be achieved by the marine plan authorities harmonizing their plans with development plans within the Local Development Framework, Shoreline Management Plans and River Basin Management Plans. The marine plan authority will be required to ensure, as far as reasonably possible, compatibility with adjacent terrestrial plans. Under the Marine and Coastal Access Act, marine plan authorities will be required to take all reasonable steps to ensure that their plans are compatible with marine plans in related plan areas. This will aid different Administrations' planning, for example, in estuaries that straddle borders.

These marine plans and the UK's regional strategies will take into consideration the River Basin Management Plans developed by the Environment Agency through the River Basin Liaison Panel. These management plans are required by the Europeam Union's Water Framework Directive, which introduced the concept of integrated river basin management.

Conclusions

BBC News reported that UK Environment Secretary Hilary Benn addressed the House of Commons in October 2009, stating that the Marine and Coastal Access Act 'provides for a streamlined regulation and better protection of marine wildlife ... In particular, it will help us identify potential conflicts arising from the fact that we put competing pressures on our seas, and find a way of doing something about it' (BBC News, 2009). But it seems the UK legislation does more than use ocean zoning to resolve conflicts – it also creates strong mechanisms to achieve the integration of land, marine and freshwater planning which is so desperately needed.

The various disparate marine spatial planning, marine protected area site selections and integrated zoning efforts can finally be brought together under the new UK legislation. Not only will this make planning more efficient and effective, it lays the groundwork for a strategic comprehensive ocean zoning apprach, should the UK government decide to move in that direction. In addition, all the analyses and assessments that have guided the marine spatial planning processes in the UK can be used to inform other zoning processes – in neighbouring countries and around the world.

References

BBC News (2009) 'Marine Bill Enters Final Stages', 11 November, available at news.bbc.co.uk/go/pr/fr/-/2/hi/science/nature/8352990.stm

Daily Telegraph (2009). 'Map of proposed marine reserves published', *Daily Telegraph*, London, published 7:01AM GMT 10 Nov 2009. Available at www.telegraph.co.uk/earth/wildlife/6530894/Map-of-proposed-marine-reserves-published.html

Department of Environment, Food and Rural Affairs (DEFRA) (2004) 'Review of marine nature conservation', Summary of working group report to government, July, available at www.defra.gov.uk/environment/biodiversity/marine/documents/rmnc-summary-0704.pdf

Department of Environment, Food and Rural Affairs (DEFRA) (2008) 'A strategy for promoting an integrated approach to coastal management in England', London

Department of Environment, Food, and Rural Affairs (DEFRA) (2009) 'Managing Marine Resources: The Marine Management Organisation', available at www.defra.gov.uk

European Commission (EC) (2007) Com (2007) 308 final – Communication from the Commission – 'Report to the European Parliament and the Council: An evaluation of Integrated Coastal Zone Management (ICZM) in Europe', available at www.eur-lex.europa.eu/LexUriServ/LexUriServ.do?uri=COM:2007:0308:FIN:EN:PDF

Scottish Natural Heritage (2009) draft document, Scottish National Heritage, Inverness, www.snh.gov.uk/planning-and-development/marine-planning

Vincent, M., A., S.M. Atkins, C.M. Lumb, N. Golding, L. Lieberknecht and M. Webster (2004) 'Marine nature conservation and sustainable development – the Irish Sea pilot'. Report to DEFRA by the Joint Nature Conservation Committee, Peterborough, UK

7
Zoning Undertaken by the OSPAR Countries of the Northeast Atlantic

www.oceanpowermagazine.net

Offshore wind farm in Danish North Sea waters

Countries in the northeast Atlantic region have been setting the pace in comprehensive ocean zoning outside marine protected areas, as we have seen in the previous chapter discussing the United Kingdom's efforts, and as will be further demonstrated in this chapter detailing other European marine spatial management initiatives.

Both individually and collectively under the regional agreement known as OSPAR, several northern European countries have adopted goals to better integrate the management of their seas, each with slightly different approaches to marine spatial planning. Belgium has amended its long-held practice of zoning on land to move spatial management out to sea. It has successfully transitioned from determining the relative biological value of different terrestrial areas to similar biological valuation of its waters, and has developed various scenarios for zoning its entire Exclusive Economic Zone. Norway has also undertaken marine spatial planning – being notable as the only European country that has thus far not shied away from encompassing fisheries management in its marine spatial planning work. In addition, both Germany and the Netherlands have planning processes in place that may well lead to comprehensive ocean zoning in their territorial waters.

The policy drivers for MSP in Europe

The Intergovernmental Oceanographic Commission publication 'Visions for a Sea Change' (Ehler and Douvere, 2007) presents a policy hierarchy for MSP that has at its highest level international policies (conventions, international agreements, directives and global policies) that in turn affect national policy development. Beyond the national policies are regional policies, such as those detailed below in the discussion of OSPAR and in HELCOM (Helsinki Commission, see Box 7.1). Whether such regional policies drive national policy, or whether instead they emerge out of national policies in a cimcumscribed region (as Ehler and Douvere seem to imply) is immaterial – it is likely that the policy conext at all scales spurs planning efforts at the national scale, which in turn inform regional and global efforts in the ever-dynamic world of marine policy development.

In the context of comprehensive ocean zoning in the northeast Atlantic region, impetus has come from the European Union (EU), the subregions such as delineated in OSPAR and HELCOM, and the international community itself. The following section, taken from Douvere and Ehler (2006), describes some of the key policies that are driving the formation of regional agreements, as well as zoning activity within key northern European countries.

> *International agreements*
> - *United Nations Convention on the Law of the Sea (UNCLOS)*
> - *Chapter 17 of Agenda 21*
> - *International Maritime Organization convention and protocols (such as MARPOL, the London Dumping Convention, Oil Pollution Preparedness Responses and Control (OPRC)*

- 1995 Global Programme of Action for the Protection of the Marine Environment from Land-Based Activities
- 1995 UN Fish Stocks Agreement, and the FAO Code of Conduct for Responsible Fisheries
- The United Nations Agreement for the Implementation of the Provisions of the United Nations Convention on the Law of the Sea of 10 December 1982 relating to the Conservation and Management of Straddling Fish Stocks and Highly Migratory Fish Stocks
- World Summit for Sustainable Development 2002
- Convention on Biological Diversity (CBD)

European legislation and initiatives
- Green Paper on the future Maritime Policy for the European oceans and seas (2006)
- EU Thematic Strategy for the Marine Environment (2005)
- EU Recommendations on Integrated Coastal Zone Management (ICZM) (2002)
- The Fifth Ministerial North Sea Conference (2002)
- European Wildlife Directives
- EU Common Fisheries Policy (2002)
- EU Water Framework Directive (2000)

More recent European Union policy initiatives with great bearing on marine spatial planning and ocean zoning in Europe include the Marine Strategy Framework Directive, launched in 2008 under the auspices of the Directorate General for the Environment (DG Environment), and the Integrated Maritime Policy under the Directorate General for Maritime Affairs and Fisheries (DG MARE) (de Santos, 2009). It is interesting to note that the EU policies can be both a driver and a constraint. The way these policies drive MSP is obvious; they can be a constraint thanks to the vagaries of EU structure in that marine planning done to meet the Marine Strategy Framework Directive cannot encompass fisheries policy, which falls under the remit of DG MARE. Norway, as is explained below, avoids this conundrum, since it is not a member of the European Union. However, Norway is part of the OSPAR Convention (described below), and its national zoning efforts have both been aided by and informed the zoning efforts of other countries in the region.

The OSPAR regional agreement

The Convention for the Protection of the Marine Environment of the North-East Atlantic (OSPAR Convention) is the mechanism by which 15 governments of the western coasts and catchments of Europe, together with the European Union, cooperate to protect the marine environment of the northeast Atlantic. It started in 1972 with the Oslo Convention against dumping and was broadened to cover land-based sources and the offshore industry by the Paris

Figure 7.1 *Map of the OSPAR area*

Source: www.ospar.org

Convention of 1974. These two conventions were unified ('OS' stemming from the Olso Convention, 'PAR' deriving from the Paris Convention), updated and extended by the 1992 OSPAR Convention. The new annex on biodiversity and ecosystems was adopted in 1998 to cover non-polluting human activities that can adversely affect the sea.

The 15 governments are Belgium, Denmark, Finland, France, Germany, Iceland, Ireland, Luxembourg, The Netherlands, Norway, Portugal, Spain, Sweden, Switzerland and the United Kingdom. Finland is not on the western coasts of Europe, but some of its rivers flow to the Barents Sea, and historically it was involved in the efforts to control the dumping of hazardous waste in the Atlantic and the North Sea. Luxembourg and Switzerland are Contracting Parties due to their location within the catchments of the River Rhine. For a map of the OSPAR area, see Figure 7.1.

Overall, the OSPAR Commission attempts to put several key principles for effective marine management into practice by guiding the member states. These principles include the ecosystem approach, the precautionary principle and the 'polluter pays' principle. In addition, OSPAR member states learn from one another and their experiences in identifying best available techniques (BAT) amd best environmental practices (BEP) for managing human impacts on the oceans.

The Biological Diversity and Ecosystems Strategy of the OSPAR Commission addresses all human activities that have the potential to impact marine biodiversity in the region, with the exception of pollution-causing activities that are covered by other strategies. Beyond that, the OSPAR Commission also targets already affected marine areas, providing guidance on their restoration (where desirable and practicable).

According to the OSPAR website (www.ospar.org), the OSPAR Commission's Strategy has four elements:

- Ecological quality objectives: In creating a pilot project on the identification of ecological quality objectives that marine management should strive towards, the Commission intends to provide practical guidance to member states in other sub-regions
- Species and habitats: On the basis of assessments of species and habitats that are threatened or in decline, OSPAR programmes are developed for their protection
- Marine protected areas: OSPAR is working towards the creation of an ecologically coherent network of well-managed marine protected areas, including in Areas in Areas Beyond National Jurisdiction
- Human activities: OSPAR works to assess human activities that may adversely affect the NE Atlantic region

In 2002 the fifth Ministerial North Sea Conference in Bergen directed the OSPAR Commission to investigate marine spatial planning as a means to resolve use conflicts in the region. It should be noted that although OSPAR targets human activities that affect the marine environment, with special focus on issues shared by the 15 member nations, the management of fisheries and the regulation of shipping are outside the Convention's mandate. However, individual member states can develop marine spatial plans that incoporate these two potentially damaging ocean uses, as we will see in the noteworthy cases of two OSPAR countries: Belgium and Norway.

Zoning within Belgium's Exclusive Economic Zone

Belgium has jurisdiction over a relatively small portion of the North Sea, with an Exclusive Economic Zone (EEZ) totalling 3600k^2. This small area is under

> **BOX 7.1 THE HELSINKI COMMISSION**
>
> A similar subregional commission, the scope of which neighbours the OSPAR region, is HELCOM, which includes all the countries that border the semi-enclosed Baltic Sea: Denmark, Estonia, Finland, Germany, Latvia, Lithuania, Poland, Russia and Sweden. The Helsinki Commission (HELCOM) is the governing body of the 'Convention on the Protection of the Marine Environment of the Baltic Sea Area', otherwise known as the Helsinki Convention. HELCOM's main focus is on controlling pollution, but marine spatial planning is beinning to crop up there as well. According to the HELCOM website (www.helcom.fi), broad-scale marine spatial planning is one of the key concepts within the Baltic Sea Action Plan that was adopted by the HELCOM member states to restore the marine environment by 2021.

intense pressure, being centrally located in one of the most heavily exploited marine areas in the world. Existing human uses of the area include shipping, oil and gas, wind energy generation, sand and gravel extraction, recreation, fishing, and military exercises; most of these uses are increasing, putting ever-greater pressure on this thin slice of North Sea continental shelf waters.

The many uses of marine resources and space in this patch of the North Sea, the increasing user conflicts and the emergence of new uses has necessitated a move away from what was previously an ad hoc approach to managing the marine environment to a forward-looking strategy fully utilizing marine spatial planning.

Belgium has been on the cutting edge of marine spatial planning, even before the term became popular in Europe and before MSP became mandated for European Union nations. In large part this pioneering ability arose from the widely accepted practice of assessing the relative value of land areas and using the derived information to guide municipal zoning, the siting of protected areas and other planning efforts. Adapting this zoning to the marine environment meant that Belgium could plan marine management with multiple objectives in mind. Comprehensive ocean zoning (COZ) within Belgium's territorial sea and Exclusive Economic Zone is a central tool of the planning effort.

Zoning in Belgium is on the one hand iconic, in that it presents a microcosm of how COZ could proceed in other marine areas with a high degree of use and heavy congestion of users. At the same time, Belgium's zoning process is unusual, since it grew organically out of the country's long-enshrined land-use planning. A key component of Belgium's land zoning has been what the country calls 'biological valuation', in which all land areas are assessed for their intrinsic biodiversity value.

Belgium has undertaken similar biological valuation for its small marine territory. This initiative has been described in detail by Derous and co-authors (Derous et al., 2007). Figure 7.2 shows the composite results of a lengthy biological valuation exercise to determine which areas need most attention for management, and which could potentially be 'sacrificed' for industrial development (gravel extraction, wind farms, etc.).

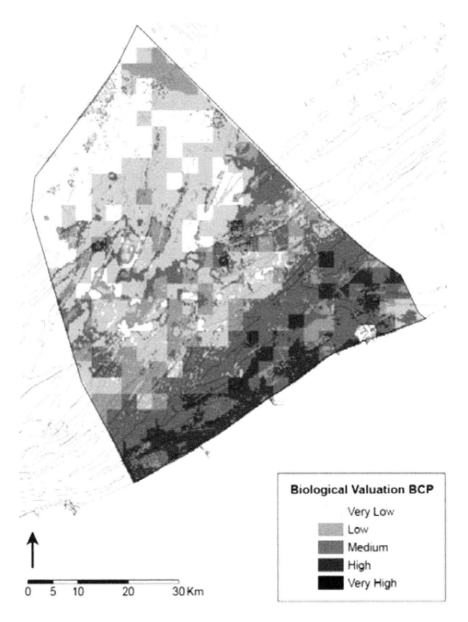

Figure 7.2 *Marine biological valuation map of the Belgian Exclusive Economic Zone*

Source: Derous, 2008

The criteria used to evaluate the relative importance of different marine areas and habitats are not unlike the criteria used to identify the location of priority areas for conservation, as in the siting of marine protected areas and reserves (see Table 7.1 for a listing of marine valuation criteria that were used in the biological valuation exercise).

Table 7.1 *Definitions of marine biological valuation criteria used by Belgium*

First order criteria

Rarity

Degree to which an area is characterized by unique, rare or distinct features (landscapes/habitats/communities/species/ecological functions/geomorphological and/or hydrological characteristics) for which no alternatives exist

Aggregation

Degree to which an area is a site where most individuals of a species are aggregated for some part of the year or a site which most individuals use for some important function in their life history or a site where some structural property or ecological process occurs with exceptionally high density

Fitness consequences

Degree to which an area is a site where the activities undertaken make a vital contribution to the fitness (=increased survival or reproduction) of the population or species present

Modifying criteria

Naturalness

Degree to which an area is pristine and characterized by native species (that is, absence of perturbation by human activities and absence of introduced or cultured species)

Proportional importance

Global importance: proportion of the global extent of a feature (habitat/seascape) or proportion of the global population of a species occurring in a certain subarea
Regional importance: proportion of the regional (e.g., NE Atlantic region) extent of a feature (habitat/seascape) or proportion of the regional population of a species occurring in a certain subarea within the study area
National importance: proportion of the national extent of a feature (habitat/seascape) or proportion of the national population of a species occurring in a certain subarea within territorial waters

Source: Adapted from Derous et al., 2006

Despite the fact that these criteria and the maps that resulted from the application of them were developed by scientific consensus, the valuation maps have not yet been adequately used to bolster a comprehensive zoning effort. In part this is because the production of the biological valuation maps came too late for incoporation in a parallel initiative that was occurring in Belgium: marine spatial planning to find the optimal location of wind farms in the country's small but heavily used EEZ (Douvere et al., 2006). Belgium had already taken some major steps in marine spatial planning (e.g., designation of some areas for wind farms and a new zoning plan for aggregate extraction) by the time the biological valuation was completed and published.

The quasi-independent drive to accommodate wind farms is notable for two reasons: 1) it illustrates the inefficiencies that result from processes that are not undertaken in a coordinated and strategic manner and 2) it shows, much like the case of zoning being spurred by wind energy development in the US state of Massachusetts (see Chapter 12), the extent to which a particular kind of industrial driver can be the impetus for ocean zoning.

Some detail on the parallel initiative mentioned above is useful. The project 'Toward a Spatial Structure Plan for Sustainable Management of the North Sea', known as the GAUFRE project, is distinct from other spatial planning efforts that were taking place in Europe at the time because it focused on methods to develop and communicate alternative scenarios for future use of ocean space. The GAUFRE project was a two-year, multidisciplinary modelling exercise done by the University of Ghent that examined present values and future goals for sea use by creating structural maps that describe various scenarios for Belgium's ocean territory (see Colour Plate 7.1 for an example). The structural maps are dynamic maps that describe a particular vision. According to the leader of the GAUFRE project, Frank Maes, they are not detailed enough to be used as a legal basis for zoning activities at sea; instead they provide guidance for future zonation or reallocation of existing activities, as far as the latter is possible (MEAM, 2008).

The GAUFRE project and the biological valuation mapping exercise have taken place in the context of ongoing efforts by Belgium to use its vast experience in land zoning to better manage ocean uses in its highly exploited and heavily stressed waters. Since 2003, Belgian authorities have been focused on meeting two particular challenges of managing ocean use: 1) determining spatial limitations for offshore wind farms and sand/gravel extraction, and 2) identifying priority areas for conservation to be proposed as candidate marine protected areas under the Natura 2000 Network (Ehler and Douvere, 2007). These two challenges spurred the development of the Belgian Master Plan, Phase I of which is focused on zonation for wind and mining industries, and Phase II of which targets Natura 2000 site selection.

The EU Directive calling for establishment of Natura 2000 sites states that damaging activities that could significantly disturb the species or deteriorate the habitats for which the site is designated should be avoided (European Commission, 2005). In addition, Natura 2000 sites chould be the focus of restoration to bring habitats and species to a 'favorable conservation status' in their natural range (Douvere and Ehler, 2008). Each member state decides how to undertake protection or restoration of a Natura 2000 site. According to Douvere and Ehler (2008), the provisions used can be statutory (e.g., making a nature reserve), contractual (e.g., signing management agreements with the land owner) or administrative (providing the necessary funds to manage the site). Whatever method is used, Natura 2000 siting must take account of the economic, social and cultural requirements and regional and local characteristics of the area concerned.

The Master Plan described above was developed in the wider context of the Belgian Marine Spatial Planning Policy Framework, which has the following core issues (Douvere and Ehler, 2008):

- Development of offshore wind farms
- Delimitation of marine protected areas

- Policy for sustainable sand and gravel extraction
- Financial resources for oil pollution prevention
- Mapping of marine habitats
- Protection of wrecks providing habitat for marine wildlife
- Management of land-based activities that affect the marine environment

However, even with such ambitious plans for managing a wide variety of uses, the current system of management in Belgium can only work through permitting, and permits are still issued on a sectoral basis. Many authors have stressed the importance of putting this multi-use planning effort in the broader context of a legally mandated management regime. Only the future will show whether some of the core issues not currently under the remit of authorities executing the Master Plan (e.g., oil pollution, biodiversity conservation through habitats assessment and mapping, management of land-based activities) will be addressed by a comprehensive ocean zoning plan, which certainly seems the logical next step.

Ocean zoning in Norway

Norway is the only country in Europe that has not shied away from encompassing fisheries management in its MSP work – ironically, this is because Norway is not in the European Union (EU) and not bound by the DG MARE Common Fisheries Policy (which, unlike the directives issued by DG Environment, are legally binding).

Norway established an 'ad hoc Nordic Forum on MPAs in Marine Spatial Planning' to disseminate knowledge about research and management, which has guided not only Norway's national planning but the use of comprehensive ocean zoning in other countries as well. Norway also created a spatial plan for ecosystem-based management of the Barents Sea and the Lofoten Islands, which was implemented in 2006 and is meant to be updated every four years. The latter provides a framework for managing oil and gas exploration and recovery, fishing and shipping, in an area extending from 1nm off the coast to the 200nm EEZ limit. Plate 7.2 shows the zoning map developed during this subregional pilot (taken from Olsen et al., 2007).

The process to develop the spatial plan for the Barents Sea and Lofoten Islands occurred in three phases (see Colour Plate 7.2). During Phase I, status reports on the condition of the marine and coastal environment of the subregion and sectoral analyses of shipping, fisheries and aquaculture industries within it were generated to characterize the subregion. During Phase II, four government-financed impact assessments were carried out to evaluate the effect of shipping, hydrocarbon extraction, fishing and external factors such as pollution on the environment, resources and local communities. These assessments provided the basis for the analyses that followed in Phase III: an appraisal of total cumulative impact by human activity, both current and projected to 2020; identification of use conflicts; setting of management goals; and highlighting of data and knowledge gaps (Olsen et al., 2007).

At the time of Phase III work, a parallel and synergistic exercise was being undertaken to develop operational environmental quality objectives (EcoQOs) for climate, ice edges, phytoplankton, zooplankton, commercial fish species, non-commercial fish species, benthic organisms, marine mammals, seabirds, select invasive or alien species, threatened species and pollutants, to be monitored annually (Olsen et al., 2007). This informed the planning process, but importantly also provides a baseline as a foundation for an adaptive management regime.

The resulting zoning plan for the subregion shows ecologically valuable areas needing strict protection and areas of existing or potential conflict requiring new or amended management. For instance, based on the information that went into developing the spatial plan, Norway requested the International Maritime Organization to move shipping lanes outside its territorial waters, in order to avoid conflicts between shipping and fisheries. Further, Norway has moved to close some areas to hydrocarbon activities and has established seasonal fishery closures. This demonstration model of ocean zoning has been invaluable, not only in improving management of the Barents Sea and Lofoten Islands, but also in providing a real life, practical example of how to do comprehensive ocean zoning.

At a workshop that the ad hoc Nordic Forum held on marine spatial planning in 2007, bringing together over 100 managers and researchers to discuss MSP and zoning in theory and in practice, participants discussed how Norway had applied zoning at subregional levels (such as the Barents Sea) to derive lessons learned (Sørensen et al., 2009). The key findings of the workshop were:

1 Effective marine spatial planning is crucial for the mitigation of evolving conflicts and further degradation of the Nordic marine environment, as well as for proactive planning to avoid future conflicts
2 Higher resolution data and mapping of marine habitats and species will create better operational tools for MSP
3 Enhanced transnational cooperation and management of maritime activities will help address large-scale cumulative impacts and help promote ecological coherence of marine protected area networks
4 The common cultural backgrounds of Nordic inhabitants, their similar languages, and the existing high level of regulation and compliance provide a solid foundation that can be built upon for transnational MSP

Marine spatial planning in other European countries

Several other northern European countries have attempted to use ocean zoning to resolve use conflicts and enhance nature conservation. The Netherlands developed a spatial planning policy framework as part of their 'Integrated Management Plan for the North Sea 2015', in order to promote the economically efficient use of their marine space. Termed the 'Dutch National Water Plan', the policy steers the Netherlands towards the creation of a safe, healthy and productive ocean (Ehler and Douvere, 2009). The initiative required the mapping of current and prospective human uses of the marine environment (oil and gas, wind energy, sand extraction and fishing, among other uses). The Dutch government then developed scenarios to ten years out (beginning in 2005 and running to 2015), looking at various levels of economic growth and positing different cost/benefit ratios for management options.

Germany is also undertaking marine spatial planning, to fulfil its obligations under both OSPAR and HELCOM, in order to meet the requirements under the various directives of the EU, and to make its national marine management more efficient and effective. In 2004 Germany extended its Federal Spatial Planning Act to its entire EEZ, extending national sectoral competencies (including MSP) and creating a legal requirement for marine spatial planning (Douvere, 2008). The draft spatial plan completed in 2007 specifies 'priority areas', 'reservation areas' and 'suitable areas' – though these sites have not received formal designation, nor are they yet part of any implemented ocean zonation. While the national level planning is ongoing, planning authorities collectively called the Länder have been engaged in extending land-use plans to the coastal environment (within the German territorial sea).

European trials with ocean zoning, in the context of marine spatial planning, nature conservation, biodiversity protection, fisheries management and conflict reduction between various users of the marine environment, have been attempted using diverse methods with varying levels of success. Most ocean zoning initiatives in Europe remain in the experimental stages and are thus works in progress. Nonetheless, lessons derived from the ocean zoning undertaken by northern European countries have already served to strengthen ocean zoning elsewhere: in the Mediterranean Sea, in Africa, across Asia and in North America.

References

Derous, S. (2008) *Marine Biological Valuation as a Decision Support Tool for Management*, Marine Biology Department, University of Ghent, available at www.vliz.be/imisdocs/publications/136055.pdf

Derous, S., T. Agardy, H. Hillewaert, K. Hostens, G. Jamieson, L. Lieberknecht, J. Mees, I. Moulaert, S. Olenin, D. Paelinckx, M. Rabaut, E. Rachor, J. Roff, E. W. M. Steinen, J. T. van der Wal, V. van Lancker, E. Verfaillie, M. Vincx, J. M. Węsławski, S. Degraer (2007) 'A concept for biological valuation in the marine environment', *Oceanologia*, vol 49, no 1, pp99–128

De Santos, E. (2009) 'Whose science – precaution and power play in European environmental decision making', *Marine Policy* (in press)

Douvere, F. (2008) 'The importance of marine spatial planning in advancing ecosystem-based sea use management', *Marine Policy*, vol 32, pp762–771

Douvere, F. and C. N. Ehler (2006) 'The international perspective: Lessons from recent European experience with marine spatial planning', available at ioc3.unesco.org/marinesp/files/Paper_SanJoseConference_FINAL.doc

Douvere, F. and C. N. Ehler (2008) 'New perspectives on sea use management: Initial findings from European experience with marine spatial planning', *Journal of Environmental Management*, vol 90, no 1, pp77–88

Douvere, F., F. Maes, A. Vanhulle and J. Schrijvers (2006) 'The role of marine spatial planning in sea use management: The Belgian case', *Marine Policy*, vol 31, no 2, pp182–191

Ehler, C. and F. Douvere (2007) 'Visions for a Sea Change', *Report of the First International Workshop on Marine Spatial Planning. Intergovernmental Oceanographic Commission and Man and the Biospehre Programme*, IOC Manual and Guides 48, ICAM Dossier no 4, UNESCO, Paris

Ehler, C. and F. Douvere (2009) 'Marine spatial planning: a step-by-step approach toward ecosystem-based management', *Intergovernmental Oceanographic Commission and Man and the Biosphere Programme*, IOC Manual and Guides 53, ICAM Dossier no 6, UNESCO, Paris

European Commission (2005) *Natura 2000, Conservation in Partnership*, European Commission, Brussels

Maes, F., J. Schrijvers and A. Vanhulle (2005) *A Flood of Space: Toward a Spatial Structural Plan for Sustainable Management of the North Sea*, Belgian Science Policy, Brussels

MEAM (2008) 'Case study: creating a zoning plan for Belgium's North Sea, with lessons from land-based zoning', *Marine Ecosystems and Management*, vol 2, no 1, pp7–8

Olsen, E. H. Gjøsæter, I. Røttingen, A. Dommasnes, P. Fossum and P. Sandberg (2007) 'The Norwegian ecosystem-based management plan for the Barents Sea', *ICES Journal*, vol 64, pp599–602

Sørensen, T. K., M. Blaesbjerg and O. Vestergaard (2009) 'Marine spatial planning in the Nordic region: perspectives, challenges, and opportunities', International Marine Conservation Congress 22–28 May 2009, Fairfax, VA, USA, Poster Presentation

8
Zoning the Asinara Marine Park of Italy

Tundi Agardy

Placid and pristine Asinaran waters

The Italian government takes good advantage of an opportunity

Off the northwestern corner of the Italian island of Sardinia lies a small, cliff-ringed island called Asinara (the name derives from the donkeys that once grazed the island's steep but lush hills). Asinara is a rarity in the overdeveloped and overexploited Mediterranean Sea – it is almost pristine, preserved intact by its designation, over a century ago, as an island penitentiary. Its status as a maximum security prison meant that not only the land remained undeveloped, but that the seas surrounding it were also preserved, through a no-entry zone that extended 10km offshore.

The opportunity to preserve a valuable and unusually pristine site as a marine protected area presented itself when the Italian government decommissioned the prison on the island in 1997. This site not only represents the least impacted coastal area in Italy, it may well be one of the least impacted coastal sites in the whole of the northern Mediterranean. When the penitentiary was finally decommissioned, the Italian government had the foresight to protect the largely undeveloped island and surrounding unscathed waters and contracted a group of consultants to develop options for the marine park. The Asinara Marine National Park was declared shortly thereafter to maximize conservation values and create opportunties for the neighbouring communities on Sardinia (see Figure 8.1).

Besides its pristineness, Asinara and its surrounding waters has high conservation value, due to its habitat-level diversity, species richness, and the presence of several threatened and rare species (e.g., the Corsican gull, the rare limpet *Patella ferruginea* and the loggerhead sea turtle *Caretta caretta*). The coastline of the island is ringed with cliffs, caves, rocky shorelines and sandy beaches, while the nearshore environment comprises muddy bottom and seagrass meadows, along with rock reefs. Fish abundance is relatively high, thanks no doubt to the decades old restricted zone surrounding the island.

Tasked to create a zoning plan

The consulting team (of which this author was a member, along with researchers from the Italian Ministry of Environment's marine research unit known as ICRAM, an economist from the University of Maryland and a fisheries biologist from the UN's Food and Agricultural Organization) decided that the designs for the marine protected areas could go in any number of directions, depending on the specific goals and objectives for management. These goals and objectives had not been stipulated when the park was established by presidential decree. (All MPAs in Italy are established through Presidential decree, with little detail offered on reasons for establishment, means of management or operational plans.)

Recognizing that the ultimate zoning plan could be arbitrary and reflect or favour the interests of a single party or stakeholder group, the team decided to use an objective method to evaluate an array of options. That is, the team

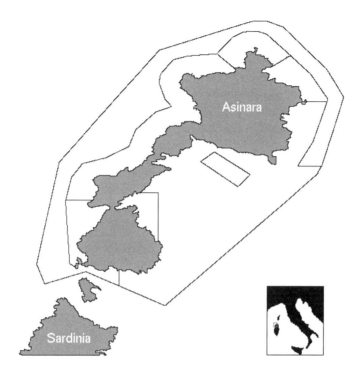

Figure 8.1 *Asinara Marine National Park, located off the northwest coast of Sardinia (Italy)*

Source: Villa et al., 2001

decided to present a set of zoning plans, showing outer boundaries of the MPA and zones within it, to assist decision-makers in evaluating trade-offs and making informed choices for the MPA design. The consulting team used Geographic Information Systems (GIS) of ecological and human use data, analysed by Spatial Multiple Criteria Analysis (SMCA), to determine which option was preferable in terms of meeting the maximum number of objectives (a form of least cost/maximum benefit analysis).

Developing the zoning options

The initial step in the developing zoning options for the Asinara National Marine Park was to undertake habitat mapping, which in European parlance is known as biocenotic mapping. Biocenotic mapping shows interacting organisms living together in a specific habitat or biotope. Colour Plate 8.1 (from ICRAM, 1999) shows a detailed map of habitat types, worked up by Leonardo Tunesi and his colleagues at the Italian scientific research agency formerly known as ICRAM (Central Institute for Scientific and Technical Marine Advice, the applied marine research body under the Italian Ministry for Environment, Land and Sea).

The systematic tool known as Spatial Multiple Criteria Analysis (SMCA) was then used to assess mapping data and derive new maps of the suitability of the potential park area to each of various protection levels. This technique, coupling GIS-based land assessment and evaluation with a formal analysis of the design priorities for each protection level, allows decision-makers to define priorities and visualize optimal zoning plans to meet priority objectives (Villa et al., 2001).

Spatial Multiple Criteria Analysis is one method among a rich and diverse set of techniques, collectively known as multicriteria evaluation, that are widely used in fields from economic analysis to environmental impact assessment. Multiple criteria evaluation methods have assisted urban and regional planning (e.g., Nijkamp et al., 1990), allowing planners to make objectively informed choices instead of simplistic common-sense decision-making. SMCA allows joint consideration of aspects as different as social preferences, development needs and conservation requirements. A fundamental technique in SMCA is concordance/discordance analysis: a set of observations, described as a set of measured attributes, is ranked according to a concordance (or discordance) score that is computed using sets of 'priority weights' expressing the importance of each attribute within a particular scenario.

According to Villa et al. (2001), it is common practice to compare use scenarios on the basis of the concordance scores computed against a description of an existing or planned situation. One of the strengths of SMCA is the ease with which heterogeneous information can be combined in the analyses. Quantitative measures can be used along with semi-quantitative information and ranks; physical attributes of an area and results of habitat surveys can be incorporated with no need for special data preprocessing.

The second step in Spatial Multiple Criteria Analysis is to define the study or target area by a set of measurable variables. The choice of variables is up to the modeller; however, the set of variables should ideally have as little internal correlation as possible, and the variables should be suitable for comparison, expressing some form of value, or level of risk (Villa et al., 2001). It should be noted that not all variables have to do with ecological integrity or other features important to conservation; cultural values can also be captured, and these are particularly important in the context of Mediterranean spatial planning (Badalamenti et al., 2000). Variables used in the Asinara zoning are presented in Table 8.1 opposite.

Deciding upon and articulating possible objectives for spatial management is the third critical step in the SMCA planning process. The outcome of this step in the process is the definition of one or more 'scenarios' – as they are called in common SMCA parlance – describing land-use priorities under each particular viewpoint through a relative importance value (weight) for each variable considered. Numeric weights or importance rankings can be assigned directly through a collaborative brainstorming process, or they can be calculated based on pairwise comparison matrices.

Given a set of variables and one or more sets of priority weights, concordance scores that describe in quantitative terms the amount of overlap of values

Table 8.1 *Variables used in the Asinara zoning exercise*

Mapped variable
Geology and geomorphology
Benthic community dominance structure
Diversity of the fish community (species counts)
Nurseries areas for fish populations
Sites relevant for the biological cycle of *Patella ferruginea*
Sites relevant for the biological cycle of *Caretta caretta*
Sites relevant in the biological cycle of marine cetaceans
Potential for transit and settling of *Foca monaca*
Hatching sites for marine fauna
Interest for archaeology based on presence of archaeological sites or knowledge of previous settlements
Interest for scientific research and education
Suitability for traditional fishing techniques, based on presence of installations or existence of traditional fishing sites
Suitability for commercial fishing
Suitability for aquaculture
Suitability for scuba diving
Suitability for snorkelling
Suitability for whale and dolphin watching
Suitability for recreational fishing
Suitability for sport fishing (underwater)
Suitability for sailing, recreational boating
Suitability for swimming
Density of commercial navigation; commercial harbours
Industrial installations; power plants
Tourist infrastructure: hotels, camping sites, etc.
Tourist harbours
Input of pollutants from urban, industrial sources and rivers
Density and severity of acoustic and other forms of pollution
Areas subjected to military control
Accessibility as a function of distance from harbours and mooring sites and terrain characteristics

and habitat features and discordance scores that describe areas with no such overlap can be computed for each different evaluation unit in the source dataset. For instance, for the purposes of strict conservation of biodiversity, MCA evaluates the degree to which different parts of the coastal environment satisfy the criteria of providing critical habitat for special species such as the rare limpet *Patella ferruginea* or the loggerhead turtle *Caretta caretta*, along with species-rich fish assemblages and nursery areas such as *Posidonia* seagrass beds. These composite values are weighted according to their surmised contribution to biodiversity conservation – e.g., habitat for the monk seal *Foca monaca* receives a lesser weight than *Posidonia* beds or habitat for *Patella ferruginea*, since the latter actually exist on the island while the former only describes potential habitat should the seal – thought to be extirpated from the area many decades ago – reappear in Asinaran waters. The concordance scores, standardized as needed, are then remapped to create a 'concordance map' for that land- or marine-use scenario, graphically portraying the agreement between the priorities specified and the features of the area under consideration. In typical applications, one concordance map is obtained for each land-use scenario. The maps can be later aggregated and analysed using standard GIS capabilities to suit the objectives.

Choosing the preferred option

Villa et al. (2001) used SMCA to evaluate the concordance between various portions of the Asinara reserve area and each of the four planned protection levels. The concordance maps were used as a guide to inform, document and justify the proposed zoning of the MPA. The concordance maps for each protection level that were used to inform the development of the actual zoning plan, as presented by Villa et al., (2001) are described below:

- Natural Value of the Marine environment (NVM). This map aggregates natural values related to the diversity and size distribution in the benthic and aquatic communities, to the presence of endemic or rare species, and to the presence and status of conservation of habitats that have crucial roles in maintaining ecosystem function (e.g., nursery areas). The map was obtained by GIS addition of properly reclassified biocenotic and habitat maps for the most important habitats and key species
- Natural Value of the Coastal environment (NVC). This map was obtained aggregating information relative to important coastal endemism, the suitability of habitats for return or reintroduction of key species (e.g., the loggerhead turtle *Caretta caretta* or the monk seal, *Foca monaca*), and the ability of the coastal habitat to support key species that nest on the mainland (such as the Corsican gull)
- Value of Area for Recreational Activities (RAV). This map was also obtained by attributing relative importance values to each variable involved and performing a SMCA to characterize the value as concordance of the area characteristics with the suitability for each feature. We considered suitability for all recreational and cultural activities as illustrated in, which also summarizes the weighting system used. The final value map was obtained from the results of SMCA after weighting with the accessibility of the area
- Value of area for Commercial exploitation of Resources (CRV). This map considered only the allowed fishing activities, related to traditional and artisan fishing practices, and was prepared by addition of maps identifying traditional fishing sites and general suitability for such practices
- Degree of accessibility of area (Ease of Access, EAC). This map was used both as a 'benefit' value for scenarios where access is allowed and encouraged, and as a 'cost' factor in high protection scenarios, being a proxy for potential disturbance. It was obtained by addition and distance buffering of maps identifying marine access routes and existing harbours

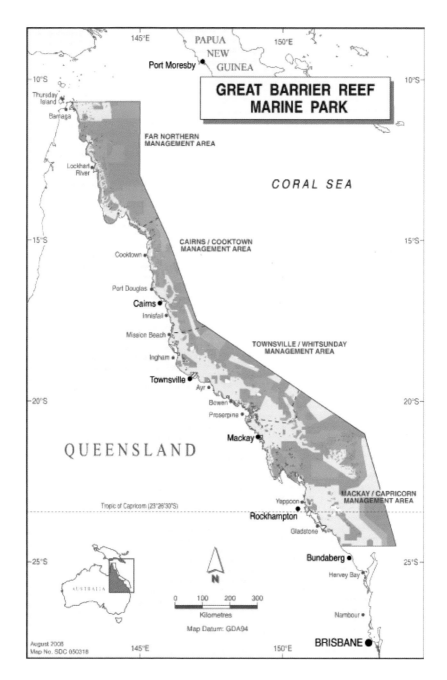

Plate **4.1** *The Great Barrier Reef Zoning Plan, showing General Use zones in light blue, Habitat Protection Zones in darker blue, Conservation Park Zones in yellow, Buffer Zones in olive, Scientific Research Zones in orange, Marine Park zones in green and Preservation Zones in pink (see Table 4.2 for details)*

Source: GBRMPA (2009)

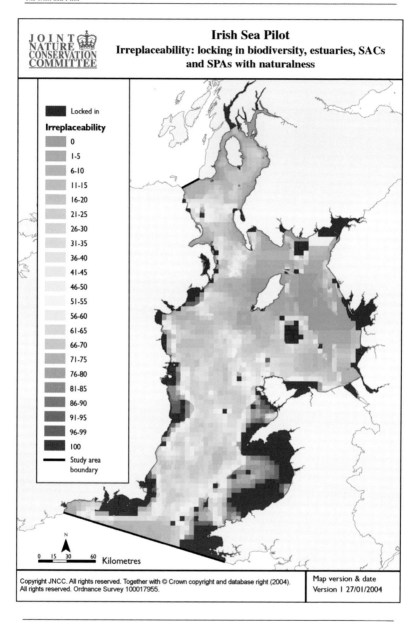

Plate 6.1 *Irish Sea Pilot irreplaceability of planning units, showing the relative importance of areas for meeting conservation targets, with existing protections 'locked in' in purple and a gradient of importance from blue-green (low) to red (high)*

Source: Vincent et al. (2004)

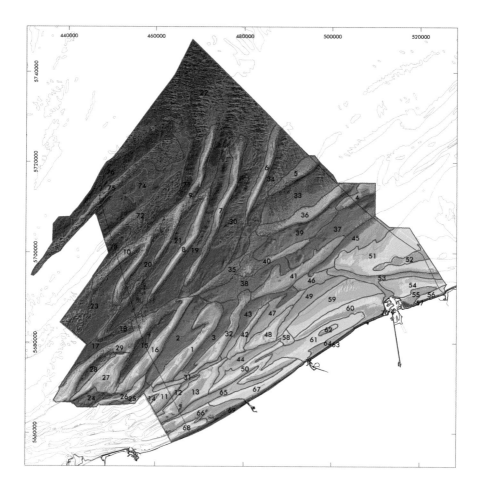
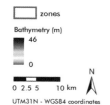

Plate 7.1 *Example of a structural zoning map produced by GAUFRE, with zones superimposed on bathymetry*

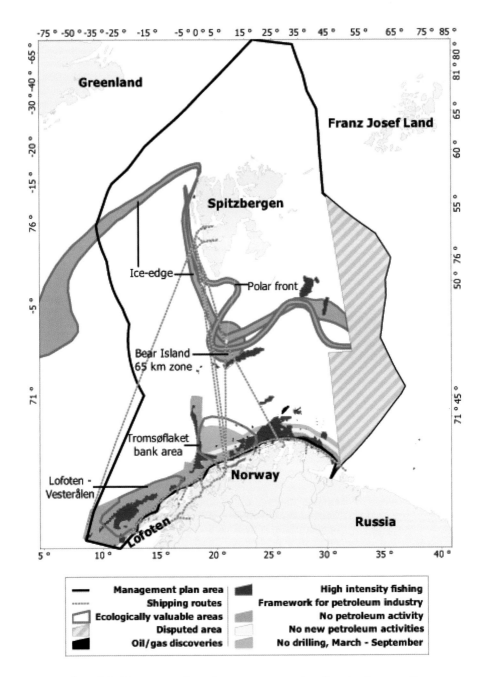

Plate 7.2 *Ecosystem-based management plan for the Barents Sea*
Source: Olsen et al. (2007)

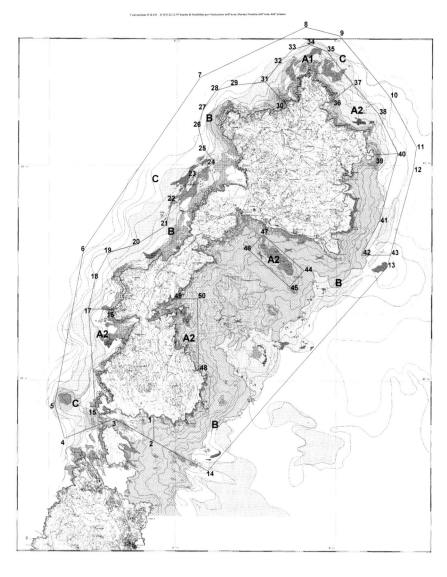

Plate 8.1 *Biocenotic (habitat) map of Asinara Island's waters, showing different biocenoses across soft bottom, hard bottom, and seagrass bed habitats shown in different colours, with tentative park boundaries and zones superimposed*

Source: ICRAM (1999)

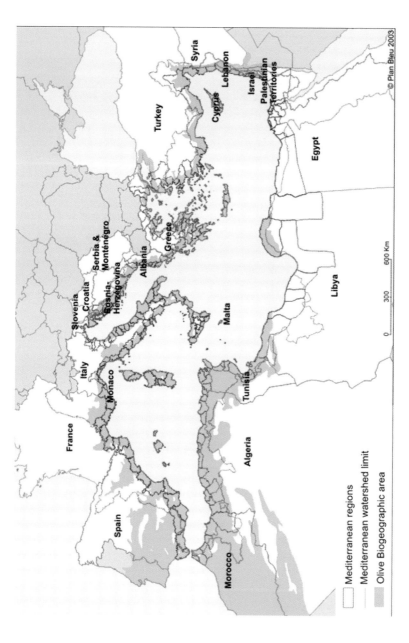

Plate 9.1 Distribution of marine protected areas in the Mediterranean Basin

Source: Abdullah et al. (2008b)

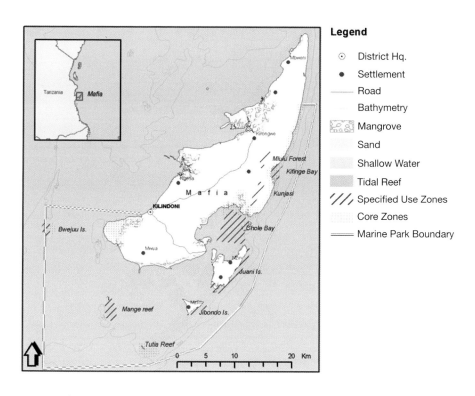

Plate 10.1 *Mafia Island Marine Park boundaries and zonation*
Source: gridnairobi.unep.org

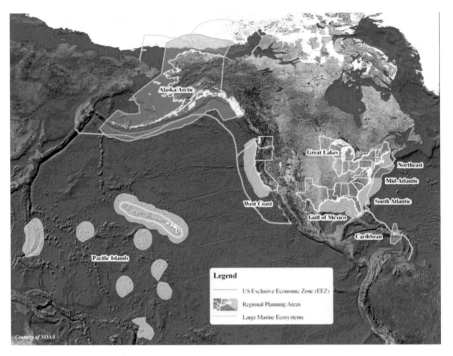

1. **Alaska/Arctic Region:** Alaska

2. **Caribbean Region:** Puerto Rico and US Virgin Islands

3. **Great Lakes Region:** Illinois, Indiana, Michigan, Minnesota, New York, Ohio, Pennsylvania and Wisconsin

4. **Gulf of Mexico Region:** Alabama, Florida, Louisiana, Mississippi and Texas

5. **Mid-Atlantic Region:** Delaware, Maryland, New Jersey, New York, Pennsylvania and Virginia

6. **Northeast Region:** Connecticut, Maine, Massachusetts, New Hampshire, Rhode Island and Vermont

7. **Pacific Islands Region:** Hawaii, Commonwealth of the Northern Mariana Islands, American Samoa and Guam

8. **South Atlantic Region:** Florida, Georgia, North Carolina and South Carolina

9. **West Coast Region:** California, Oregon and Washington

Plate 12.1 *Nine Regional Planning Areas proposed for the US and corresponding minimum state representation*

Source: CEQ (2009)

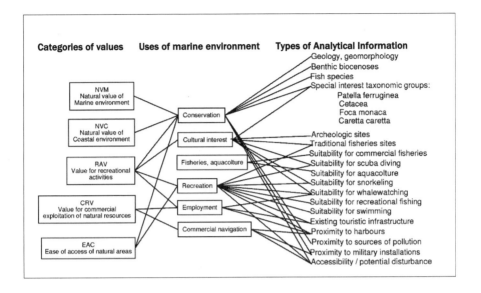

Figure 8.2 *Structuring of information used in the Asinara MPA zoning process*

Source: Villa et al., 2001

While many of these values overlap in targeting objectives such as conservation, preservation of cultural sites or recreation, some of the values could be mutually exclusive (commercial exploitation and endangered species protection, for instance). Figure 8.2 describes how these value maps relate to various categories of objectives and to data or information about the marine environment and its prospective uses.

The resulting concordance maps shown in Figure 8.3 provide the necessary foundation for developing a variety of zoning plans. The results provide a range of area options for each of the four zones employed in Italy's standard system of marine zonation within marine protected areas: strictly protected A Zones, subdivided into entry and no-entry areas; regulated use B Zones; and surrounding areas inside the MPA (C Zones). Computer-driven GIS allows the planner to pick optimal configuration to target desired objectives.

In Villa et al. (2001), running the known values for different portions of the target area or habitats within it against the management objectives for the MPA allowed the Multiple Criteria Analysis to produce several possible zoning plans. The optimal zoning plan that best accommodates all uses, as based on the results of concordance mapping, shown in Figure 8.1, with A1 and A2 Zones being the most strictly protected, B Zones buffering the A zones and a General Use C Zone.

In addition to developing options for zonation, the consulting team considered the types of management/regulations for each of the zones, which would allow objectives and goals for each zoning scenario or option to be

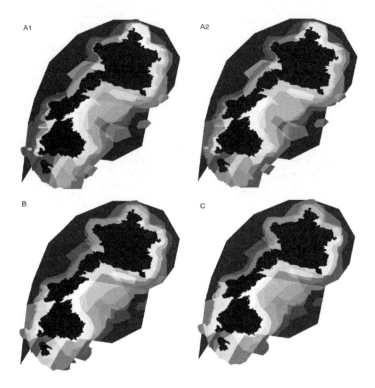

Figure 8.3 *Concordance maps for the protection scenarios in the Asinara MPA. Concordance levels range from black (minimum) to white (maximum) and are scaled on 256 levels. Areas in white thus represent best areas for each of the four types of zones, depending on what values are optimized*

Source: Villa et al., 2001

met. In doing so, the planners considered all current and existing uses, and determined which uses would be appropriate where, and at what level of use. Table 8.2 opposite lists the permitted activities for each zone.

Summary

Asinara National Marine Park is considered a model for MPA planners who need to evaluate values at proposed MPA sites and optimize management for various targeted objectives. In fact, this approach has already been employed at other Italian MPAs that were decreed but not yet zoned. The vast majority of Italian MPAs have only simplistic zoning, based on pre-existing use patterns, so the potential for rezoning is vast in the Italian MPA system. However, this tool can only be used in information-rich situations – otherwise value ranking and weighting becomes too subjective and the power of objectivity is lost.

Table 8.2 *Permitted and prohibited activities in each zone, where A1 represents strictly protected, no-go zones; A2 represents access but no-take zones; B is a buffer zone of regulated activity; and C is the general zone encompassing all of the protected area not within A or B zones. For each activity and protection level the letter P stands for prohibited, R for subject to specific limitations, A for allowed upon written authorization, and G for generally allowed.*

Category	Activity	A1	A2	B	C
Research	Non-destructive monitoring	A	A	G	G
Sea access	Sailing	P	R	G	G
	Motor boating	P	P	R	R
	Swimming	P	P	G	G
Staying	Anchoring	P	P	R	R
	Mooring	P	R	A	G
Recreation	Diving	P	R	A	G
	Guided tours	P	R	A	G
	Recreational fishing	P	P	R	G
Exploitation	Artisanal fishing	P	P	R	R
	Sport fishing	P	P	P	R
	Scuba fishing	P	P	P	P
	Commercial fishing	P	P	P	P

Yet the proof is in the pudding; whether the Asinara National Marine Park or any other Italian MPA meets its goals and objectives will depend not only on the quality of the chosen zoning plan, but also on the extent to which active management – stimulating compliance with regulations and having adequate monitoring, surveillance and enforcement – is really put into place. As has been made evident by (Guidetti et al. 2008), MPA establishment without active management (particularly enforcement) has, not surprisingly, not led to measurable ecological outcomes. Considerations of enforcement practicality will have to be considered in all ocean zoning plans, lest such plans suffer the fate of the world's considerable number of meaningless 'paper parks' (see especially Kareiva, 2008).

Regardless of the efficacy of management in this small protected area, the Asinara planning process provides a demonstration of useful techniques for ocean zoning at wider scales. Perhaps the most useful lesson learned from this case study is that tying zoning plans to specific objectives allows decision-makers to evaluate trade-offs and make informed choices, and allows the public to understand the rationale for the ways that ocean uses and access are regulated. But scaling up such an effort to the level required for comprehensive national zoning, or regional zoning such as will be discussed in the next chapter, remains an enormous challenge.

References

Badalamenti, F., A. A. Ramos, E. Voultsiadou, J. L. Sanchez Lizaso, G. D'Anna, C. Pipitone, J. Mas, J. A. Ruiz Fernandez, D. Whitmarsh and S. Riggio (2000) 'Cultural and socio-economic impacts of Mediterranean marine protected areas', *Environmental Conservation*, vol 27, pp110–125

Guidetti, P., M. Milazzo, S. Bussotti, A. Molinari, M. Murenu, A. Pais, N. Spanò, R. Balzano, T. Agardy, F. Boero, G. Carrada, R. Cattaneo-Vietti, A. Cau, R. Chemello, S. Greco, A. Manganaro, G. Notarbartolo di Sciara, G. F. Russo and L. Tunesi (2008) 'Italian marine protected area effectiveness: Does enforcement matter?', *Biological Conservation*, vol 141, pp699–709

ICRAM (1999) *Studio di fattibilità per l'istituzione dell'area marina protetta dell'Isola dell'Asinara prevista dall'articolo 36 della Legge Quadro sulle aree protette N 394/91*, Ministero dell'Ambiente, Rome, 3 volumes

Kareiva, P. (2006) 'Conservation biology: beyond marine protected areas', *Current Biology*, vol 16, no 14, R533–535

Nijkamp P., P. Rietveld and H. Voogd (1990) *Multicriteria Evaluation in Physical Planning*, Holland Publishers, Amsterdam

Villa, F., L. Tunesi and T. Agardy (2001) 'Optimal zoning of a marine protected area: the case of the Asinara National Marine Reserve of Italy', *Conservation Biology*, vol 16, no 2, pp515–526

9
Possibilities for Holistic Zoning of the Mediterranean Sea

Tundi Agardy

Dropoff near Tavolara, Sardinia – an example of the many highly productive and diverse areas of the Mediterranean Sea

The highly valued, highly stressed Mediterranean Sea

Among the world's oceans and seas, the Mediterranean vies for the distinction of having the longest history of intensive use and cultural importance. But ironically, this semi-enclosed sea is at once highly treasured and undervalued. While both ancient civilizations and modern societies have acknowledged the sea's importance, large swaths of the basin remain unmanaged and open to threats.

The Mediterranean Sea comprises a vast set of ecosystems which are diverse, productive and which have not been systematically studied, especially those parts of the sea away from the nearshore environments of European coastal countries. Superimposed on this complicated backdrop of interconnected and threatened marine systems is a rapidly evolving legal framework for managing access and use. Twenty-nine nations have watersheds that drain into the sea (Figure 9.1), and most with a Mediterranean coastline claim a territorial sea stretching to 12nm offshore. Beyond territorial jurisdictions are the high seas and what is known as ABNJs (Areas Beyond National Jurisdictions), which currently constitute the bulk of the Mediterranean Basin's volume of 2.5 million square kilometres. The jurisdictional picture of the Mediterranean is changing, however, as countries begin to expand their claims to 200nm offshore, utilizing designations such as Exclusive Economic Zones, Conservation Zones and other extended jurisdictions.

Despite the rapidly changing legal framework, there are several multilateral initiatives which either use marine spatial planning directly or lay the groundwork for eventual MSP and ocean zoning at the regional level. Under the aegis of the Barcelona Convention and its Regional Activity Centre for Specially Protected Areas (RAC/SPA), protected areas in the ABNJs could not only preserve the integrity of this globally important region, but also provide the basis for eventual ocean zoning. The 1995 Specially Protected Areas protocol of the Barcelona Convention provides for Specially Protected Areas of Mediterranean Importance (SPAMI) listings, which confer international distinction on certain areas.

Neither a representative network of marine protected areas nor a system of zoning has been elaborated for the Mediterranean, though under the aegis of the Mediterranean Regional Seas Programme's RAC/SPA several initiatives have laid the groundwork for Mediterranean spatial planning processes. As is shown in Colour Plate 9.1, at last count there were 57 marine protected areas scattered around the sea, predominantly in the northern and western portions of the basin, and almost exclusively within a scant few kilometres of the land (Abdullah et al., 2008a; 2008b). The Pelagos Sanctuary for Mediterranean Marine Mammals described in Box 9.1 at the end of this chapter is a trilaterally established marine protected area in the Ligurian Sea which includes territorial waters of Italy, Monaco and France, as well as areas beyond national jurisdiction. It is the one and only high seas protected area in the Mediterranean and is considered by some the first high seas marine protected area in the world (Notarbartolo di Sciara et al., 2008).

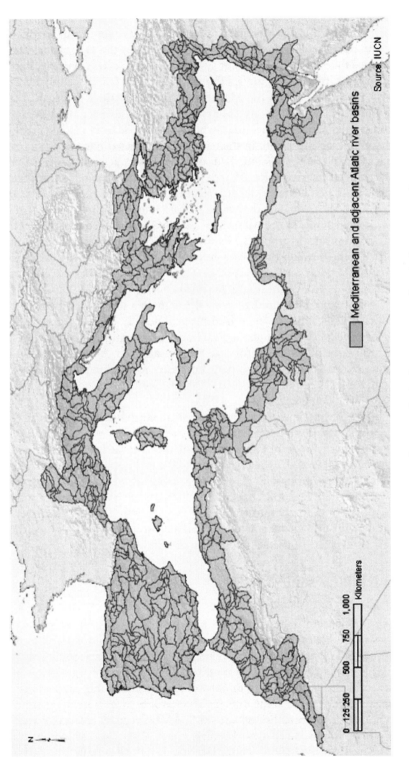

Figure 9.1 Countries with watersheds draining into the Mediterranean Sea

Source: Cuttelod et al., 2008

The current distribution of marine protected areas would not seem to be a sufficient basis for larger scale ocean zoning, given the geographic and biophysical gaps that exist in the current system of priority sites for protection. However, there are precedents for spatial planning that could well result in comprehensive ocean zoning of the Mediterranean sea. At the Basin scale, the SPAMI planning initiative could lay the groundwork for an integrated marine spatial plan for Mediterranean high seas, which could well be coupled to spatial planning efforts occurring within in the territories of coastal nations.

However, ocean zoning in some form or another has actually taken place beyond the realm of marine protected area planning in the Mediterranean. For instance, the International Maritime Organization of the United Nations (IMO) recently agreed to shift some of the world's busiest shipping lines away from dolphin foraging grounds in the Alboran Sea (westernmost Mediterranean off the coasts of Spain and Morocco). Historically, large numbers of merchant ships sailed straight through the area east of the Strait of Gibraltar, near the Spanish city of Almeria and the Cabo de Gata National Park, which also includes offshore marine protected areas. Amazingly, nearly 30 per cent of the world's maritime traffic used to pass through this designated protected area, disrupting cetaceans on their feeding grounds. According to the UK newspaper *The Independent* (Popham, 2007), the decision by the IMO, whose implementation is being carefully followed, came after Dr Ana Canadas and her colleague Ricardo Sagarminaga spent five years researching shipping and conservation conflicts for the European Commission's LIFE-Nature project (www.europa.eu/environment/life).

Other policy developments in the Mediterranean have had the de facto effect of creating certain kinds of ocean zones as well. The regional fisheries authority known as the General Fisheries Commission for the Mediterranean or GFCM (formed in 1949 under the Provisions of the UN Food and Agricultural Organization's Constitution, with 23 member states, as well as the European Community) declared in 2008 that all Mediterranean waters deeper than 1000m were off limits to bottom trawling (www.gfcm.org). This effectively creates a no-trawl zone – what some would call a protected area – spanning the bulk of the vast expanse of the Mediterranean's seafloor.

Planning a representative network of protected areas for ABNJ

Recognizing the need to better represent under-represented geographies and habitat types in the Mediterranean-wide system of marine protected areas, the UNEP Regional Seas Programme in the region spearheaded a project to identify priority marine sites in Areas Beyond National Jurisdiction. The intent of the project was to identify ecologically important and/or vulnerable places in the Mediterranean high seas, or areas outside of territories, which member states could then nominate to protect select sites as SPAMI (Specially Protected Areas of Mediterranean Importance) sites. The first phase of the process (completed in 2009) focused on identifying candidate areas, which entailed collating information, creating a database and undertaking objective analyses to identify specific ABNJ areas meriting further attention (Notarbartolo di Sciara and Agardy, 2009). Criteria for this site selection process were adapted from other site selection methodologies, including the Convention on Biological Diversity (CBD) criteria for creating representative networks of marine protected areas to protect biodiversity, and the 'Azores criteria' specifically designed to steer site selection in the high seas (CBD, 2008). The project took place under the auspices of the Barcelona Convention's Regional Activity Centre for Specially Protected Areas, based in Tunis.

Phase 1 work entailed three steps. The steps were designed to span a hierarchy of scales; the first step in the process was the biogeographical assessment that allowed the identification of ecologically coherent subregions within the Mediterranean Basin. The steering committee for the project agreed that subdividing the Mediterranean Sea into subregions would help to ensure that any eventual MPA network arising from the site selection process would be truly representative of all regions, as well as all habitat types. While previous researchers had divided the Mediterranean either into two large subregions (East and West) or seven smaller subregions (see Spalding et al., 2007), this project enumerated eight distinct subregions: Alboran Sea, Algero-Provençal Basin, Tyrrhenian Sea, Adriatic Sea, Ionian Sea, Tunisian Plateau – Gulf of Sidra, Aegean Sea and Levantine Sea.

The second step in the process was to review existing criteria, adapt them and add additional discriminating features to guide the selection of sites, specific to the consideration of Mediterranean high seas and offshore regions. A review of existing information/databases revealed that data are patchy and information quality is inconsistent across tax and geographical regions. Nonetheless, enough information was present to guide the application of region-specific criteria (relying heavily on the CBD criteria that emerged from the Azores meeting, with additional criteria to guarantee that the resulting network would conserve biological diversity and ecological integrity to the maximum extent possible), and supplement the expert-opinion driven selection of priority sites undertaken in the third step of the process.

New guidelines for identifying areas to designate as protected areas were developed during an 'Expert workshop on ecological criteria and biogeographic classification systems for marine areas in need of protection' organized in the Azores in October 2007 under the auspices of CBD (BD, 2008). This meeting produced a set of criteria to guide site selection; these criteria were later adopted during the CBD's thirteenth Subsidiary Body on Scientific, Technical and Technological Advice Meeting held in Rome in February 2008. In particular, the Azores workshop produced:

1 Scientific criteria for identifying Ecologically or Biologically Significant Areas ('EBSAs') in need of protection, in open-ocean waters and deep-sea habitats, including examples of features that would meet such criteria
2 Scientific criteria and guidance for selecting areas to establish a representative network of marine protected areas, including in open-ocean waters and deep-sea habitats

Table 9.1 *Criteria used in identifying Ecologically or Biologically Significant Areas*

- Uniqueness or rarity (to the best of the available knowledge)
- Special importance for life history stages of species
- Importance for threatened, endangered or declining species and/or habitats
- Vulnerability, fragility, sensitivity or slow recovery
- Biological productivity
- Biological diversity
- Naturalness

Table 9.2 *Oceanic features considered in the selection of EBSAs*

Benthic features

- Seamount communities
- Cold water coral reefs
- Coral, sponge and bryozoan aggregations
- Hydrothermal vent ecosystems
- Cold seeps
- Canyons
- Trenches

Pelagic habitats

- Upwelling areas and areas of high productivity
- Fronts
- Gyres

Critical habitats and corridors for vulnerable or highly migratory species

- Whales and other cetaceans
- Seabirds
- Sea turtles
- Sharks
- Highly migratory fish
- Discrete deep-sea fish populations

Source: Notarbartolo di Sciara and Agardy, 2009

Table 9.1 lists the categories of criteria commonly used in identifying EBSAs (CBD 2008).

A non-exhaustive list of examples of features that would meet the above criteria for identifying ecologically or Biologically Significant Areas (or species), provided in the CBD workshop report, include many features that are relevant to ocean zoning planning considerations as well. These are shown in Table 9.2.

The workshop also provided a useful set of guidelines for the selection of areas to establish a representative network of MPAs, including in open-ocean waters and deep-sea habitats. Starting from an overarching goal of a global representative network of MPAs ('Maintain, protect and conserve global marine biodiversity through conservation and protection of its components in a biogeographically representative network of ecologically coherent sites'), the Azores meeting suggested that the coherence of such a network 'can be attained by diverse mechanisms that promote the genetic flow, through connectivity, among populations of marine organisms with planktonic life history phases. Amongst others are ocean currents providing homogeneity within a dispersal area and geographical distance and barriers that promote isolation and associated biological diversity' (CBD, 2008).

There are also useful criteria developed for Particularly Sensitive Sea Areas (PSSAs) under the IMO (International Maritime Organization, 2006) which assisted in the Mediterranean effort and could be useful to zoning methodologies used elsewhere. These are provided in Table 9.3.

For highlighting particularly important sites for biodiversity, data on benthic invertebrates, fish fauna, sharks, birds, marine turtles, pinnipeds and cetaceans were particularly useful to the Mediterranean ABNJ site selection process. In addition, information on key biogenic and physical habitats in the ABNJ domain provided a useful baseline for the hierarchical methodology that begins with the largest regional scale, moves through subregions (or ecoregions) and further focuses on key areas within those, either to identify potential MPA sites or potential zones in an ocean zoning process.

The third and final step in the site selection process comprising Phase 1 is the selection of priority areas within each of the eight Mediterranean ecoregions. This itself entails three discrete steps, again utilizing a hierarchical process taking the planners from larger scales to smaller: 1) identifying the priority regions (EBSAs) in each of the Mediterranean ABNJs subdivisions using the refined site selection criteria, 2) applying feasibility criteria to the previously highlighted priority areas in order to identify potential sites that could be protected as SPAMIs, and 3) preparing a short list of potential sites in the ABNJs which could be protected as SPAMIs.

In order to select areas of conservation significance or concern EBSAs within which potential SPAMI sites were elaborated), the RAC/SPA project surveyed key experts regarding various aspects of Mediterranean ecology and marine biodiversity. This expert-opinion or Delphic process was used because regional and subregional databases were inconsistent, with large information gaps in areas of the eastern and southern reaches of the Mediterranean Sea. Experts were asked to highlight especially important areas within each subregion, using standardized criteria which they then ranked according to the extent to which it helped them in their determination (Notarbartolo di Sciara and Agardy, 2009).

The numerous resulting polygons were overlaid to highlight especially critical areas from an ecological or biodiversity perspective. Manual concordance mapping resulted in 10 EBSAs being identified in all. Within each of these EBSAs, collating expert opinion and existing information allowed the identification of smaller sites, resulting in a listing of 15 highest priority areas, within which RAC/SPA and the parties to the Barcelona Convention could develop SPAMI nominations in the future.

Table 9.3 *IMO Criteria for Identifying Particularly Sensitive Sea Areas*

Ecological criteria
4.4.1 Uniqueness or rarity – An area or ecosystem is unique if it is 'the only one of its kind'. Habitats of rare, threatened, or endangered species that occur only in one area are an example. An area or ecosystem is rare if it only occurs in a few locations or has been seriously depleted across its range. An ecosystem may extend beyond country borders, assuming regional or international significance. Nurseries or certain feeding, breeding, or spawning areas may also be rare or unique.
4.4.2 Critical habitat – A sea area that may be essential for the survival, function, or recovery of fish stocks or rare or endangered marine species, or for the support of large marine ecosystems.
4.4.3 Dependency – An area where ecological processes are highly dependent on biotically structured systems (e.g., coral reefs, kelp forests, mangrove forests, seagrass beds). Such ecosystems often have high diversity, which is dependent on the structuring organisms. Dependency also embraces the migratory routes of fish, reptiles, birds, mammals, and invertebrates.
4.4.4 Representativeness – An area that is an outstanding and illustrative example of specific biodiversity, ecosystems, ecological or physiographic processes, or community or habitat types or other natural characteristics.
4.4.5 Diversity – An area that may have an exceptional variety of species or genetic diversity or includes highly varied ecosystems, habitats, and communities.
4.4.6 Productivity – An area that has a particularly high rate of natural biological production. Such productivity is the net result of biological and physical processes which result in an increase in biomass in areas such as oceanic fronts, upwelling areas and some gyres.
4.4.7 Spawning or breeding grounds – An area that may be a critical spawning or breeding ground or nursery area for marine species which may spend the rest of their life-cycle elsewhere, or is recognized as migratory routes for fish, reptiles, birds, mammals, or invertebrates.
4.4.8 Naturalness – An area that has experienced a relative lack of human-induced disturbance or degradation.
Integrity – An area that is a biologically functional unit, an effective, self-sustaining ecological entity.
Fragility – An area that is highly susceptible to degradation by natural events or by the activities of people. Biotic communities associated with coastal habitats may have a low tolerance to changes in environmental conditions, or they may exist close to the limits of their tolerance (e.g., water temperature, salinity, turbidity or depth). Such communities may suffer natural stresses such as storms or other natural conditions (e.g., circulation patterns) that concentrate harmful substances in water or sediments, low flushing rates, and/or oxygen depletion. Additional stress may be caused by human influences such as pollution and changes in salinity.
Biogeographic importance – An area that either contains rare biogeographic qualities or is representative of a biogeographic 'type' or types, or contains unique or unusual biological, chemical, physical, or geological features.

Social, cultural and economic criteria
Social or economic dependency – An area where the environmental quality and the use of living marine resources are of particular social or economic importance, including fishing, recreation, tourism, and the livelihoods of people who depend on access to the area.
Human dependency – An area that is of particular importance for the support of traditional subsistence or food production activities or for the protection of the cultural resources of the local human populations.
Cultural heritage – An area that is of particular importance because of the presence of significant historical and archaeological sites.

Scientific and educational criteria
Research – An area that has high scientific interest.
Baseline for monitoring studies – An area that provides suitable baseline conditions with regard to biota or environmental characteristics, because it has not had substantial perturbations or has been in such a state for a long period of time such that it is considered to be in a natural or near-natural condition.
Education – An area that offers an exceptional opportunity to demonstrate particular natural phenomena.

Will adding up protected zones and MPAs lead to comprehensive ocean zoning?

The consultants to the RAC/SPA process detailed above also elaborated a road map for carrying this to the eventual development of an ecological and representative network of marine protected areas using SPAMI designations in ABNJ areas (Notarbartolo di Sciara and Agardy, 2009). Next steps include a threat and socioeconomic factors analysis in order not only to identify vulnerable sites needing protection as SPAMIs, but also to be able to factor in feasibility.

Parties to the Barcelona Convention and Focal Points to the RAC/SPA will be considering these and other subsequent initiatives. Whatever is decided, any further phase of the effort should have three essential components: 1) development of a strategic plan to elaborate the priorities within the SPAMI list, 2) targeted research to determine with greater specificity the ecological characteristics of each priority area, its boundaries and direct threats to the biodiversity the area supports, and 3) analyses to determine the optimal spatial management scheme for each of the SPAMIs, including whether protected areas should be zoned, what sort of regulations should be instituted, how areas should be monitored and regulations enforced, and the appropriate governance regime for these ABNJ areas (Notarbartolo di Sciara and Agardy, 2009). These results should help guide RAC/SPA in presenting possible options for the Contracting Parties to the Barcelona Convention to consider in future SPAMI designations, in order to protect Mediterranean marine biodiversity.

Beyond SPAMI designations, there is great potential in the Mediterranean, as in other offshore areas, to establish seasonal or dynamic zones to protect highly migratory species and open ocean ecological processes. Hyrenbach et al. (2000) discuss design constraints in terms of creating dynamic protected areas – but the lessons apply equally well to establishing dynamic zones within a ocean zoning regime.

Foundations for zoning are being laid at every possible scale in the Mediterranean region, including the establishment of small-scale MPAs such as Asinara Marine National Park in Italy (see Chapter 8), Port-Cros Marine Park in France or any number of other multiple use protected areas around the basin. At larger geographic scales, the Pelagos Sanctuary for the Protection of Mediterranean Marine Mammals (Box 9.1), the GFCM no-trawl zone at depths greater than 1000m and the selection of sites that conservation groups feel should be closed to fishing in order to protect key species (such as the World Wildlife Fund petition to close areas around the Balearic Islands to promote recovery of the bluefin tuna; see WWF, 2007) could all be important pieces in a comprehensive zoning process. However, as in the case of ocean zoning in New Zealand (Chapter 5), no one can predict whether these important steps, seemingly taken towards a systematic and strategic approach to ocean management, will end up resulting in a comprehensive ocean zoning plan for this highly used and valued sea.

Box 9.1 The Pelagos Sanctuary for Mediterranean Marine Mammals

The Pelagos Sanctuary for Mediterranean Marine Mammals is a vast marine protected area extending over 87,500km^2 of sea surface in a portion of the northwestern Mediterranean Sea comprised between southeastern France, Monaco, northwestern Italy and northern Sardinia, and encompassing Corsica and the Tuscan Archipelago.

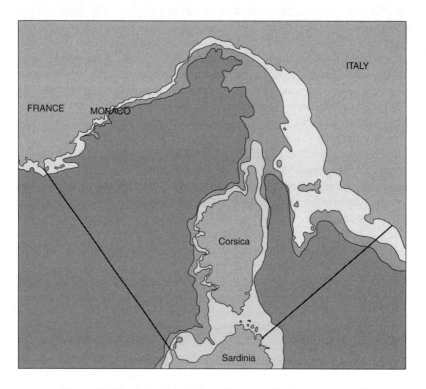

Figure 9.2 *The limits of the Palagos Sanctuary for the protection of marine mammals in the Ligurian Sea*

Source: Courtesy of Consorzio Mediterraneo, Rome

The sanctuary waters include the Ligurian Sea and parts of the Corsican and Tyrrhenian Seas, and contain the internal maritime (15 per cent) and territorial waters (32 per cent) of France, Monaco and Italy, as well as the adjacent high seas (53 per cent). Coastlines bordering on the sanctuary are predominantly rocky, with the exception of eastern Corsica and Tuscany, where they are mostly flat. Accordingly, within the sanctuary area the continental shelf is wide only in correspondence of such limited coastal plains, whereas it is mostly narrow and disseminated with steep, deeply cut submarine canyons elsewhere. The western offshore portion of the sanctuary consists of a uniform abyssal plain 2500–2700m deep. However east of Corsica the sea bottom is shallower (1600–1700m) and uneven. Compared to the rest of the Mediterranean, this marine area is characterized by very high levels of offshore primary productivity, caused by the interplay of oceanographic, climatic and geomorphological factors. A dominant cyclonic current, flowing north along Corsica and Tuscany and thence hugging the coast of Liguria and mainland France in a westerly direction, creates a permanent frontal system which acts as a boundary between coastal and offshore waters. Intense biological activity is generated along this boundary by the dynamics of the water masses associated with the front. Such phenomena are seasonally and intermittently reinforced by vertical mixing and coastal upwellings, generated by the prevailing northwesterly wind ('mistral'), which lift up from the deep waters into the euphotic zone nutrients and organic substances contributed by rivers, most notably the Rhone. Consequent high levels of primary production, with chlorophyll concentrations exceeding $10g/m^3$, support a conspicuous biomass of highly diversified zooplankton fauna, including gelatinous macrozooplankton and swarming euphausiid crustaceans (krill), *Meganyctiphanes norvegica*. Zooplankton, in turn, attracts to the area various levels of predators, mammals included.

The sanctuary contains habitat suitable for the breeding and feeding needs of the entire complement of cetacean species regularly found in the Mediterranean Sea; these include fin whales *Balaenoptera physalus*, sperm whales *Physeter macrocephalus*, Cuvier's beaked whales *Ziphius cavirostris*, long-finned pilot whales *Globicephala melas*, Risso's dolphins *Grampus griseus*, common bottlenose dolphins *Tursiops truncatus*, striped dolphins *Stenella coeruleoalba*, and short-beaked common dolphins *Delphinus delphis*.

Such remarkable cetacean faunal diversity must coexists in the sanctuary with very high levels of human pressure. The greater part of the coastal areas bordering on the sanctuary, particularly on the mainland, is heavily populated and disseminated with large and medium-sized coastal cities, ports of major commercial and military importance, and industrial areas. Furthermore, the entire sanctuary coastal zone contains important tourist destinations, thereby subject to considerable added human pressure during the summer months. As a consequence, a range of diverse human activities exerts several actual and potential threats to cetacean populations in the sanctuary, including: habitat degradation, regression and loss caused by urban, tourist, industrial and agricultural development, also associated with pollutant input in correspondence of the larger agglomerates and river mouths; disturbance from intense maritime traffic (e.g., from passenger, cargo, military, fishing and pleasure crafts), particularly high in summer, as well as from a burgeoning whale watching industry, from military exercises, research activities at sea, and from seismic oil and gas exploration; a growing risk of collisions with vessels, also in connection with the increase of high-speed passenger transportation and with offshore motorboat racing contexts; and mortality caused by the accidental entanglement in pelagic driftnets, which continue to be used in the area in spite of a ban on driftnetting imposed on fleets of European member states.

Stimulated by the awareness of the importance of the Ligurian Sea and adjacent waters for cetaceans, the Tethys Research Institute launched a project in 1990 to establish a marine protected area in the high seas, encompassing the most important habitat for cetaceans in the region. The rationale behind the proposal, which was named 'Project Pelagos', included: the ecological representativeness of the area, its high species diversity, its intense biological activity, the presence of critical habitat for a number of pelagic species including cetaceans and the opportunities that the area offered to baseline research. Most importantly, the proposal openly intended to challenge the mainstream legal notion of the time that establishing a protected area in the high seas was impossible.

Source: From Agardy and Staub, 2005

References

Abdullah, A., M. Gomei, D. Hyrenbach, G. Notarbartolo-di-Sciara and T. Agardy (2008a) 'Challenges facing a network of representative marine protected areas in the Mediterranean: prioritizing the protection of underrepresented habitats', *Ices (International Council for the Exploration of the Seas) Journal of Marine Science*, Advance Access published 28 October 2008, no 66, 7pp.

Abdullah A., M. Gomei, E. Maison and C. Piante (2008b) *Status of Marine Protected Areas in the Mediterranean Sea*, IUCN, Malaga, and WWF, France, 152pp

Agardy, T. and F. Staub (2005) 'Marine Protected Areas', A module developed for the Center for Biodiversity Conservation, American Museum of Natural History, New York

Convention on Biological Diversity (CBD) (2008) 'Report on the expert workshop on ecological criteria and biogeographic classification systems for marine areas in need of protection', SBSTTA 13th Meeting, Rome, 18–22 February 2008, 25p

Cuttelod, A., N. Garcia, D. Abdul Malak, H. Temple and V. Katariya (2008) 'The Mediterranean: A biodiversity hotspot under threat', in J.-C. Vié, C. Hilton-Taylor and S. N. Stuart (eds) *The 2008 Review of the IUCN Red List of Threatened Species*, IUCN Gland, Switzerland

Hyrenbach, K. D., K. A. Forney and P. K. Dayton (2000) 'Marine protected areas and ocean basin management', *Aquatic Conservation: Marine and Freshwater Ecosystems*, vol 10, pp435–458

International Maritime Organization (2006) 'Revised guidelines for the identification and designation of Particularly Sensitive Sea Areas', Resolution A.982(24) adopted on 1 December 2005 (Agenda item 11), 13pp

Notarbartolo di Sciara, G. and T. Agardy (2009) 'Draft interim report to the RAC/SPA: Identification of potential SPAMIs in Mediterranean Areas Beyond National Jurisdiction', Contract no 01/2008_RAC/SPA, High Seas

Notarbartolo di Sciara, G., T. Agardy, D. Hyrenbach, T. Scovazzi and P. van Klaveren (2008) 'The Pelagos Sanctuary for Mediterranean marine mammals', *Aquatic Conservation: Marine and Freshwater Ecosystems*, vol 18, pp367–391, DOI: 10.1002/aqc.855

Popham, P. (2007) 'Shipping lanes moved to boost dolphin numbers', *The Independent*, 24 April, available at www.independent.co.uk/news/world/europe/shipping-lanes-moved-to-boost-dolphin-numbers-445970.html

Spalding M. D., H. Fox, G. Allen, N. Davidson, Z. Ferdaña, M. Finlayson, B. Halpern, M. Jorge, A. Lombana, S. Lourie, K. Martin, E. MCManus, J. Molnar, C. Recchia and J. Robertson (2007) 'Marine ecoregions of the world: a bioregionalization of coastal and shelf areas', *Bioscience*, vol 57, no 7, pp573–583

WWF (2007) 'On the brink: Mediterranean bluefin tuna – the consequences of collapse', World Wide Fund for Nature, 11pp

10
Integrated Coastal Management, MPA Networks and Large Marine Ecosystems in Africa

Tundi Agardy

Fisherman poling through Mafia Island's shallows, Tanzania

Biosphere reserves and MPA networks in West Africa

West Africa's coastal zone is strategically important to the development of all seven countries of the RAMPAO (Réseau Regional d'Aires Marines Protégées en Afrique de l'Ouest) subregion (Mauritania, Senegal, The Gambia, Guinea Bissau, Cape Verde, Guinea and Sierra Leone), supporting nearly 15 million inhabitants (see Figure 10.1). The marine ecosystems, which include offshore upwelling areas, shallow banks and nearshore estuaries, are highly productive. Yet like many other areas around the world, coastal and marine systems here have been compromised in recent years due to excessive use and/or poor regulation in fisheries, tourism, development of oil and gas industry, and other ocean and coastal uses. It appears that traditional rules and conventional management have been insufficient in maintaining the fragile equilibrium, given globalization, increasing population and the push for economic development.

Figure 10.1 *West African countries in the RAMPAO region*
Source: MEAM, 2008a

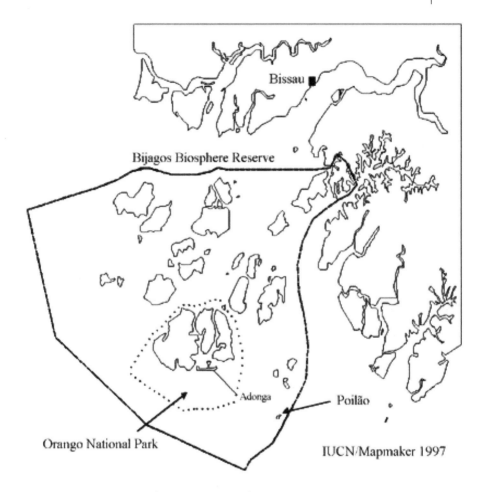

Figure 10.2 *The Bijagos Biosphere Reserve in Guinea Bissau*

Bijagos Biosphere Reserve

One of the earliest uses of spatial management in the West African region is the Bijagos Biosphere Reserve, spanning the Bijagos archipelago in Guinea Bissau, site of Africa's largest contnental shelf area (Figure 10.2). The Biosphere Reserve uses zoning to protect ecologically and cultural important areas, including core, buffer, restoration and transition area designations.

The Biopshere Reserve covers 20 major inhabited islands, 26 smaller and only transiently inhabited islands, and 37 uninhabited islets (Agardy, 1997). The marine waters of the reserve are home to marine mammals as well as a wide diversity of fish, mollusks and crustaceans, and provide critical habitat for sea turtles and a wide array of avifauna. Several tens of thousands of people live permanently in the Bijagos – they are either of Bijagos descent (an indigenous people), descendants of Nyominka fishermen who originated in Senegal or immigrants to the islands.

The Bijagos Biosphere Reserve was planned as a way to safeguard both the diverse and productive shelf waters, and the traditional cultures of the island people. Core areas, therefore, were designated as much to protect key ecological processes and hotspots for wildlife, as they were to protect sacred sites or other areas important to maintaining the culture and traditions of Bijagos people. Buffer areas surrounding the strictly protected cores were designed to allow low-level exploitation for subsistence, but at the same time stave off threats from commerical exploitation, shipping and large-scale development (Agardy, 1997). The biosphere reserve also includes areas marked for their restoration potential.

The Bijagos Biosphere Reserve has provided West African governments and civil society with a working model for sustainable development and marine management. In the hopes of replicating the success of the Guinea Bissau effort, and establish an ecologically coherent network of protected areas to conserve the whole of west Africa's continental shelf ecosystems, the West African regional marine protected areas network (RAMPAO) was established.

RAMPAO

In recent years, governmental and non-governmental organizations (NGOs) from the West African subregion that runs from Mauritania to Sierra Leone have recognized that the existing problems need to be addressed at a regional scale if the structure and the functions of the marine and coastal ecosystems are to be conserved at a regional scale (MEAM, 2008b).

Recognizing the common problems, concerns for migratory species, shared resources, mobility of users, governmental and non-governmental bodies throughout West Africa see a continuing need to address coastal zone and marine resource management at the ecoregional level. With the help of the organizations listed in Table 10.1, RAMPAO initiative was launched in 2001 in order to develop a regional network of marine protected areas (MPAs).

The RAMPAO network has a dual nature: physically connecting ecologically or socio-culturally critical sites, and linking people and institutions through a human network for exchanges, mutual reinforcement and economies of scale. It has involved all types of stakeholders (government administration, local and international NGOs, research, fishermen associations, local communities) and has political support from each of the seven countries. Importantly, the network's regional objectives take into consideration concrete interests and needs at national and local levels (MEAM, 2008b).

The RAMPAO project aims to improve the partnerships and co-management mechanisms with local communities by promoting participatory mechanisms and community-management. MPAs are considered one of the main tools for biodiversity conservation and sustainable fisheries management by the RAMPAO partners, but they are also considered important for sustainable development. The involvement of local communities in MPA management demonstrates this, as they not only preserve their cultural heritage but also provide reference sites for a better understanding of ecosystem functioning at very large scales.

Table 10.1 *International organizations involved in developing the regional strategy*

Subregional Fisheries Commission – CSRP
Regional Network for Coastal Planning in West Africa
United Nations Educational, Scientific and Cultural Organization – UNESCO
RAMSAR Convention Secretariat
World Conservation Union – IUCN
Worldwide Fund for Nature – WWF
International Foundation for the Banc d'Arguin – FIBA
Wetlands International – WI
Centre National de la Recherche Scientifique/Géomer (France)
Office National de la Chasse et de la Faune Sauvage – ONCFS (France)
Institut de Recherche pour le Développement – IRD (France)
Fisheries Information and Analysis System Project – SIAP

Since the project goes well beyond standard MPA-based management, it is fast becoming a bona fide EBM initiative. Included under the framework of the programme are fisheries management, species and habitat conservation, sustainable tourism, scientific research, and education and outreach.

The RAMPAO strategy document details how coastal and marine management at the site level contributes to more effective management at the regional scale. Each marine protected area operates at a local level, where the participation and support of local communities and stakeholders is absolutely imperative. Impacts (both positive and negative) are more immediately felt at this level where people are often asked to make significant investments of time and space. The success or failure of any protected area depends on adapting management approaches to their unique socioeconomic and natural environment to resolve problems that they themselves have identified. At the same time, conservation must operate within a context defined by national policy and legislation.

In order to be more functional, the regional MPA network requires the creation of new MPAs to fill remaining gaps. Work is now under way to evaluate what is currently missing from the network, in terms of representativeness of all habitat types, connectivity between MPAs and sites of high ecological value, and adequate management capacity. Once this gap analysis is completed, we can provide guidance on where to site new MPAs and how to reinforce the management effectiveness of existing ones.

Studies carried out at regional level on the distribution, abundance and ecology of important coastal and marine resources are being used to identify areas in need of protection or management from the 'bigger picture' perspective. This is particularly important for the management of migratory species. Regional level analyses of environmental monitoring data can also help establish whether changes observed at particular sites are simply a local phenomenon or reflections of larger, more widespread environmental changes. Other activities of mutual interest like professional training, exchange visits and preparation for international conferences can also benefit from coordination at the regional level.

As clearly articulated in the strategy, multilevel cooperation can also open the door to special fundraising opportunities – a high priority for everyone. Many bilateral and multilateral agencies as well as private foundations recognize the importance of coordinating environmental management on a larger scale as a key component in the war against poverty and as a way to preserve many of the region's globally outstanding natural wonders. The RAMPAO network of marine protected areas is designed to respect local needs while integrating national and regional priorities. With this effort, the region's institutions have created a unique opportunity for the international community to support economic and environmental development at all levels simultaneously (www.rampao.org).

Important lessons about cross-scale ecosystem-based management (ERM) that have emerged even in the early stages of this programme (provided by Charlotte Karibouye, cited in MEAM, 2008b):

- Wide (and lengthy) consultations are essential for defining clear and shared objectives and for the definition and adoption of site selection criteria
- Involving all key actors in the whole process is crucial (managers, local communities and resource users representatives, NGOs, research institutions, technical bodies, administration)
- Political support from the states was decisive (general policy declaration signed by ten ministries from six countries)
- The network's (regional) objectives must take into consideration the concrete needs at local the site (local/national) level(s)
- The (human) network is a formidable opportunity for mutual learning and reinforcement
- The technical and financial support from international partners was essential

There is every reason to anticipate the full implementation of the RAMPAO programme, and other EBM spin-offs. The Canary Current Large Marine Ecosystem Project (CCLME) started in early 2008, allowing the integration of RAMPAO in its activities. The CCLME provides a critical component of financial support to the process of building capacity to implement the RAMPAO network's plan of action.

There is a shared vision for EBM in West Africa: conservation and management of the biological and cultural diversity and of marine and coastal ecosystems of the West African coastal zone through concerted initiatives promoted in the framework of regional partnership. That there is strong political support to this regional strategy can be seen in the general policy declarations signed in 2003 by ten ministries from six countries – not only ministries in charge of protected areas but those of fisheries and the environment as well. Entities participating in RAMPAO could easily move to the next step: building on the nwtork of MPAs to develop ocean zoning plans for the entire region – zoning plans that are jointly developed by countries and insitutions all sharing a common vision.

ICM and protected areas in Tanzania

Across the African continent, the East African countries have also been experimenting with ocean zoning. Coastal zoning of one form or another has been tried in Kenya, Tanzania, Mozambique and Madagascar (as well as in South Africa to the south). However, Tanzania has been a trendsetter in coastal and marine management, through national policy and legislation, as well as through the establishment of iconic marine parks as well as a funding mechanism to support management within them (and beyond them). Some of these initiatives utilize ocean zoning; others could be seen as providing an important foundation for national-level comprehensive ocean zoning.

Mafia Island Marine Park

Mafia Island and its chain of small islets lie approximately 120km south of Dar es Salaam and 20km off the mainland, near the easternmost portion of the Rufiji Delta. The Rufiji Delta, with its vast mangrove forests, is one of the largest delta ecosystems in Africa and provides critical fish nursery grounds for much of eastern Africa (Agardy, 1997).

The main island of Mafia supports a population of over 40,000 people (Hurd, 2004) – most of whom are dependent on marine resources in the fishing, boat-building and ecotourism industries.

The island complex comprises a variety of tropical marine habitats including coral reefs, seagrass beds, mangrove forests and inter-tidal flats, as well as a remnant of threatened lowland coastal forest, about half of which is inside park boundaries. The marine biodiversity is exceptionally high, with high fish species diversity, two sea turtle species that nest on the park's beaches and dugongs. In addition to its natural wealth, Mafia Island is also home to numerous historic ruins, some dating back to the thirteenth century.

The Resolution of the National Assembly declaring the Mafia Island Marine Park (MIMP) was passed on 27 April 1995 with effect from 1 July that same year; the MIMP boundary was officially gazetted 6 September 1996 (Hurd, 2004). Mafia Island Marine Park was the first marine park designated in Tanzania, and thus required the passage of a Marine Reserves Act in Parliament in order to become officially recognized.

The MIMP is a multiple-use marine park (IUCN classification VI – see Annex) covering 822km^2, over 75 per cent of which is marine (Hurd, 2004). The MIMP covers the southern part of Mafia Island and includes the inhabited islands of Chole, Juani, Jibondo and Bwejuu as well as several other uninhabited islets.

The MIMP General Management Plan (GMP) was adopted by the Minister of Natural Resources and Tourism (MNRT) in September 2000. The purposes of the MIMP include:

- Conservation of biodiversity and ecosystem processes
- Sustainable resource use, rehabilitation of damaged ecosystems
- Involvement of local residents in development and management

Table 10.2 *The Mafia Island Marine Park Zonation Scheme*

Zone	Protection Level	Criteria for Selection	Resource Use
Core	high	Areas that warrant priority conservation status and local users can afford to wholly relinquish	Extractive use prohibited; controlled tourism and scientific use permitted
Specified Use	intermediate	Areas that warrant priority conservation status but local users not able to relinquish	Significantly destructive uses prohibited; extractive uses permitted by residents of the park
General Use	low	All other areas	Controlled extractive use permitted; priority to residents, but non-residents also permitted with authorization
Buffer	very low	800m outwards from boundary	Developments subject to EIA under MPR Act

Note: EIA = environmental impact assessment; MPR = Marine Protected Resources.

- Stimulation of rational development of underutilized natural resources
- Promotion of environmental education
- Research and monitoring of resource conditions and uses
- Conservation of historic monuments, ruins and cultural resources
- Facilitation of appropriate ecotourism development

The MIMP utilizes a zoning scheme consisting of core, specified use and general use zones, surrounded by a buffer zone. The levels of protection, criteria for selection and allowable uses are provided in Table 10.2. Colour Plate 10.1 shows the zoning scheme for the MIMP.

According to Hurd (2004), the overall picture of the park's long term sustainability is fairly positive, but still uncertain due to reliance on the tourism industry to generate revenues. One potential source of revenues that is not currently being tapped at all by MIMP is licensing and taxation of marine products. The Mafia District Council (MDC) has the legal mandate to collect these fees and is doing so but there is great potential for streamlining the collection system and increasing revenues. Thus, over the next few years, MIMP will need to continue to benefit from external donor support (Hurd, 2004). However, the Deep Sea Fishing Authority (DSFA) Act Amendments Bill that was passed by the Parliament in February of 2007 contains agreements on revenue sharing and other previously contested issues relating to proceeds from the highly valued commercial fisheries of Tanzania's 200nm Exclusive Economic Zone. This is in line with a goal of the Tanzania Marine and Coastal Environmental Management Project to establish and support a common and ecologically sustainable governance regime for offshore fisheries management. One of its features is the Marine Legacy Fund, which will generate financial support for sustainable development and conservation through revenues generated from fisheries licensing fees and biodiversity offsets from other sectors such as oil and gas (see World Bank, 2005). This complements Tanzania's already existing Coastal Village Fund, which supports alternative livelihood ventures. Plans for the Marine Legacy Fund, when fully implemented, will allow the government to derive management revenues from the licensing of commercial fishing fleets – some of these funds could be used to support the management of the park, including monitoring and surveillance to see the extent of compliance with zoning regulations.

Chumbe Island Coral Park

Chumbe Island is situated 13km southwest of Zanzibar Town (on the island of Zanzibar to the north of Mafia Island) and covers an area of approximately 20 hectares. The island is uninhabited and has no permanent structures but for the facilities built to accommodate tourists. The island was gazetted in 1994 as a protected area by the government of Zanzibar (a quasi-independent government associated with the United Republic of Tanzania). A private company was created for the management of Chumbe, known as the Chumbe Island Coral Park Ltd, by the owners of an ecotourism resort on the island.

According to the resort owners, the objectives of the Chumbe Island Coral Park (CHICOP) project are non-commercial, while operations follow commercial principles (CSI, 2006). The overall aim of CHICOP is to create a model of sustainable conservation area management where ecotourism supports conservation and education. Profits from the tourism operations are to be re-invested in conservation area management and free island excursions for local schoolchildren.

About two-thirds of the investment costs of approximately US$1 million were financed privately by the project initiator (a conservationist and former manager of donor-funded aid projects). Several project components, such as the construction of the visitors centre, biological baseline surveys, the Aders' duikers sanctuary, the park rangers' patrol boats and nature trails received some funding from donors, e.g., GTZ-GATE, GTZ-EM, the German Tropical Forest Stamp programme, EC-Microprojects, the Netherlands Embassies in Kenya and Tanzania, the WWF-Tanzania and the International School Schloss Buchhof, Munich, among others. This covered about a third of the investment costs. More than 30 volunteers from several countries provided, and continue to provide, crucial professional support for one month to three years at a time. Running costs are mostly covered from income generated through ecotourism. Seven 2-bed eco-bungalows offer accommodation for up to 14 guests. In addition, day trips are offered for up to 12 visitors.

The resort owners claim there are clear long-term benefits when the private sector establishes and manages small marine parks, as seen in resource protection, environmental awareness and economics. Overfished and depleted reefs adjacent to and upstream of the marine park are being restocked, and local people and tourists are educated about related issues. Private management is considerably less costly and more efficient than government-controlled management bodies set up by overfunded donor projects. The overall Chumbe project receives no donor or other support and depends entirely on income from ecotourism (CSI, 2006).

Integrated coastal management at the national level

The United Republic of Tanzania is moving towards an ecosystem approach by simultaneously scaling up from local initiatives, and scaling down from national governance (MEAM, 2008c). The core pieces that allow the forging of connections include national coastal zone planning, a new fisheries management Act, a nascent MPA network initiative and a sustainable development/poverty alleviation strategy. All these broad-scale initiatives have explicit links to management at the community or site-level, and many utilize spatial planning and even ocean zoning.

A report entitled 'Blueprint 2050' presented a vision for EBM in the United Republic of Tanzania – a flexible framework for overall marine management in mainland Tanzania and Zanzibar, identifying possible priorities for marine protected area networks, fisheries management and sustainable development projects (World Bank, 2005). 'Blueprint 2050' explicitly tackles the question of scale by discussing connectivity, comprehensiveness and adequacy, and by linking coastal and marine planning to sustainable financing and alternative livelihood projects at the community level. The presentation of this vision, in plain terms and in a storytelling format, has garnered understanding and support for EBM at the national level.

As Tanzania's marine management programme continues moving forward, it will keep an eye to an even bigger scale: that of the western Indian Ocean (MEAM, 2008c). Projects are already under way to explore transboundary cooperative management of coastal and marine areas shared by Kenya to the north and Mozambique to the south. And the country is a key player in larger scale regional initiatives, including SWIOFP, the Southwest Indian Ocean Fisheries Project, and WIOMSA, the Western Indian Ocean Marine Science Association.

Namibian coastal policy and its role in the Benguela large marine ecosystem

That South Africa has experimented with ocean zoning has already been stated. But another country stands out in the southern African region: Namibia. In previous writings I have ventured to call Namibian efforts among the most outstanding worldwide in terms of providing examples of how ecosystem-based management can be achieved (Agardy, 2009). Namibia is addressing the issues of coastal access and use, large-scale versus artisanal fisheries, the management of human activities in areas far offshore (and within the Benguela Current system – one of the most productive marine ecosystems in the world), and how to use marine protected areas and spatial management to achieve conservation objectives. All of this is happening in the context of nascent self-rule: Namibia only achieved its independence some 20 years ago, and much of its citizenry is involved in planning, in being government watchdogs (with a very strong anti-corruption campaign) and in working out new governance arrangements at local, regional and national levels.

There are four important pillars to Namibia's exemplary management. These are 1) the newly declared Namibian Islands MPA, 2) two recently linked coastal national parks, 3) the Benguela Large Marine Ecosystem programme and 4) Namibia's emerging coastal policy.

The recently declared Namibian Islands MPA signalled the country's commitment to adopting an ecosystem approach to fisheries management. The MPA is Namibia's first, spanning almost 1 million hectares of islands and important ocean habitat for birds, rock lobster, finfish and marine mammals. But rather than being a simple marine reserve, the MPA accommodates many different uses, including commercial and recreational fisheries, ecotourism and even diamond mining. This cross-sectoral management has meant that even though the protected area was designed with seabird protection and fisheries management in mind, the reach of management extends beyond fisheries to a wide variety of uses. Ocean zoning is one of the key tools in meeting these multiple objectives, and the development of zoning plans involved players from the line ministries, regional and local authorities, the private sector and conservation NGOs.

Nearly simultaneously, the Sperrgebiet National Park and the Namib Naukluft National Park were proclaimed – two coastal parks that span the entire coastline of the country. With cooperative planning and an institutional structure that allows for cooperative management, the new park designations create a land-sea link to promote co-management between the Ministry of Environment and Tourism, the Ministry of Fisheries and Marine Resources and regional and local authorities, in order to work towards a common objective. Figure 10.3 shows the extent of the coastal parks, as well as the Namibian Islands MPA.

Farther offshore is Namibia's portion of the Benguela Current. This upwelling area, spanning portions of the Exclsuive Economic Zone of Namibia, Angola and South Africa, as well as some of the high seas beyond, is considered by some the most productive marine system in the world. Efforts to manage impacts on this valuable area began decades ago; in 1995 the Benguela Current Large Marine Ecosystem Project was begun to address problems shared by the three countries, including the management and migration of valuable fish stocks across national boundaries, harmful algal blooms, alien invasive species and transboundary pollutants. The formation of the Benguela Current Commission (BCC) enables the three countries to engage constructively but in a voluntary fashion; by 2011, it is expected that a legally binding mechanism will be established under the Commission to resolve marine management issues.

Finally, there is the emerging coastal policy for Namibia, which was recently (25 July 2009) released to the public in the form of a green paper. Significant stakeholder engagement went into developing the policy options, and additional public feedback is expected concerning prioritization of issues, institutional structure for coastal management and other issues of appropriate governance.

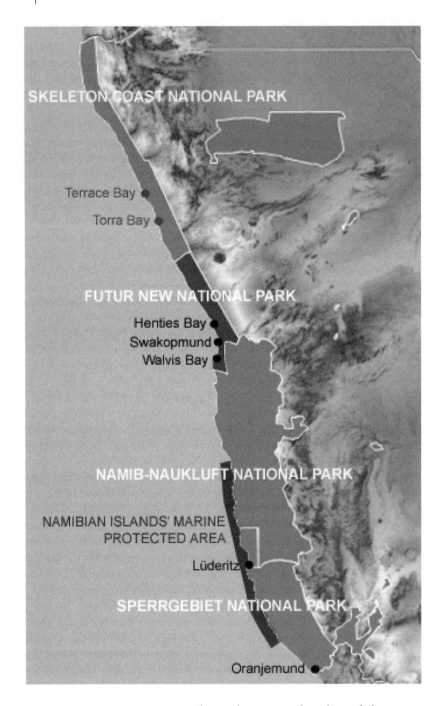

Figure 10.3 *Location of Namibian coastal parks and the Namibian Islands MPA*

Source: Courtesy of NACOMA

Namibia's coastal and marine policies thus reflect all the central tenets of ecosystem-based management, including participatory planning, integrated (cross-sectoral) management, an attempt to find the best institutional structure, strong feedback loops between science and policy, and mechanisms for adaptive management.

Even the way Namibia defines it coastal zone predisposes it to be able to be strategic about marine management. Both socioeconomic and ecological criteria are used to determine the landward extent of the coastal zone, including the communities affecting the ocean environment and affected by it and the geographic area covered by persistent sea fog. Seaward, the focal area of management includes the Namibian portion of the Benguela LME – currently to 200nm offshore (but potentially even greater if Namibia wins its bid in the UN to extend its EEZ to the outer edge of the continental shelf, some 300nm off the coast).

Ocean zoning is utilized to varying degrees in these developments in Namibia. It remains to be seen whether this country can demonstrate true integration across these four realms of marine management activity, using a common zoning framework as the 'glue' to bring these efforts together in a whole greater than the sum of its parts. Even more interesting will be to see if Namibia, or any other African country, moves towards wholesale ocean zoning throughout its jurisdiction, or across entire regions, as a way to effectively manage ocean uses and guarentee sustainability.

Summary

The challenges of economic development, poverty reduction and conflict resolution coexist with the need to conserve biodiversity and make resource use sustainable across the entire African continent and its myriad coastal and marine ecosystems. Different African countries and regions handle these challenges differently as they come into the marine management arena with a wide range of capacities for management. But even with limited capacity (and even more limited economic resources to devote to the problem), many of the marine spatial planning initiatives undertaken by African countries offer key lessons that the rest of the world would do well to heed. Whether in the design of marine protected area networks, as is being spearheaded by the RAMPAO Secretariat, or in finding ways to sustainably finance multiple use marine parks, such as has happened in Tanzania, or the comprehensive approach to protecting coastlines, offshore islands, and large marine ecosystems that Namibia has taken, dedication and innovation have been the trademarks of the African efforts to apply zoning in a wide variety of circumstances.

References

Agardy, T. (1997) *Marine Protected Areas and Ocean Conservation*, R.E. Landes Press, Austin, TX, USA

Agardy, T. (2009) 'Ecosystem-based management: The Namibian way', *Marine Ecosystems and Management*, vol 3, no 1, p5

CSI (2006) www.csiwisepractices.org/?read=185
Hurd, A. K. (2004) *Sustainable Financing of Marine Protected Areas in Tanzania*, World Bank, Washington, DC
MEAM (2007) 'Case study: connecting scales of management for EBM in West Africa', *Marine Ecosystems and Management*, vol 1, no 2, pp3–5
MEAM (2008a) *Marine Ecosystems and Management*, vol 1, no 2, web-based supplemental information at www.meam.net
MEAM (2008b) 'Case study: Connecting scales of management in West Africa', *Marine Ecosystems and Management*, vol 1, no 2, pp4–5
MEAM (2008c) 'Issues of scale: Ensuring that EBM works at all levels, from local to national and beyond', *Marine Ecosystems and Management*, vol 1, no 2, p2
World Bank (2005) 'Blueprint 2050', Washington, DC

11
Zoning within Marine Management Initiatives in British Columbia, Canada

Tundi Agardy

Vancouver Island seascape, near Campbell Island, British Columbia

A spate of marine spatial planning initiatives in British Columbia

At the time of researching and writing this book, the province of British Columbia in western Canada could arguably claim the distinction of having more spatial planning efforts going on simultaneously than any other place in the world. The work being done in these initiatives has been top notch and has contributed to a much fuller understanding of the nature of the marine ecosystems in the region, their value and prospects for their management. However, time will tell whether the planning efforts will have been worthwhile, for two reasons: 1) there appears to be little information exchange between the initiatives, resulting in a fair amount of duplication, redundancy and inefficiency, and 2) it is not clear that any of the planning efforts will actually lead to ocean zoning or, in fact, the kinds of changes in human behaviour that are needed to effect better management. As the province of British Columbia faces ever-increasing demands for resources and space, in a climate-changing future, the need for having all this effort translate into comprehensive ocean zoning plans may well increase.

Biogeography of the British Columbia coast

British Columbia spans the southern extent of Canada's Pacific coast – a vast region with nearly 30,000km of coastline, over 6000 islands, and habitats ranging from deep fjords to shallow mudflats, estuaries, kelp forests, seagrass meadows and productive fishing banks. The biodiversity of this region is surprisingly high for a non-tropical system: over 6500 species of marine invertebrates, 400 species of fish, 161 species of marine birds and 29 species of marine mammals. The waters of British Columbia have been classified as belonging to 11 ecoregions or ecosections, including 5 distinct straits, fjords, sounds, shelf area, continental slope, a transitional Pacific region and a subarctic Pacific region (see Figure 11.1).

The marine ecosystems of British Columbia are highly valued by the province's inhabitants. For the indigenous First Nations peoples, coastal areas are vital for the food they provide, but also for ceremonial and spiritual reasons. The Inside Passage (comprising the numerous straits) is one of the most popular cruising, sailing and kayaking destinations in the world. In a survey of scuba divers, British Columbia's coast was rated as the best overall destination in North America, outranking destinations such as the Florida Keys, the Gulf of Mexico and southern California. The Pacific coast of British Columbia is estimated to contribute up to CAN$4 billion annually to the nation's economy, and one in every three dollars spent on tourism – a substantial economic sector in the province – goes toward marine or marine-related activities (Living Oceans Society website: www.livingoceans.org).

Some of these ecosystem values are protected by the province through a system of protected areas, but the emphasis of this protection has been very focused on the nearshore environment. Recognizing this, various consortia

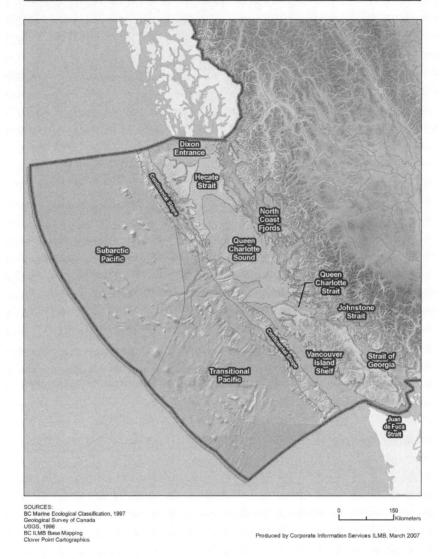

Figure 11.1 *Map of British Columbia's waters, showing ecosections*
Source: Integrated Land Management Bureau (ILMB), 2007

of government agencies and institutes (including First Nations governments), environmental non-governmental organizations and academic institutions have launched initiatives to better understand the biogeography of British Columbia's marine environments, explore the connectivity between various areas and the

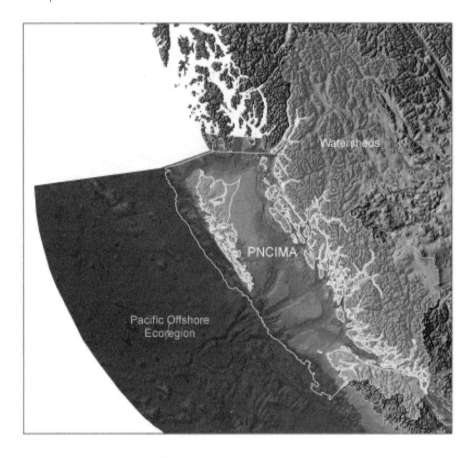

Figure 11.2 *Limits of the PNCIMA region*
Source: David Suzuki Foundation

marine habitats they contain, and determine what uses are appropriate where. Taken together, these initiatives could constitute one of the most sophisticated planning efforts for ocean zoning known anywhere. However, at this point in time, many of these initiatives operate independently, and there has been no effort to tackle marine spatial planning in a systematic and strategic fashion.

Spatial planning through various initiatives

PNCIMA
Spatial planning initiatives have been occurring at various scales in British Columbian waters. At the largest scale, the Department of Fisheries and Oceans (DFO) of Canada has been engaged in identifying priority ocean areas for conservation, as stipulated under the 1996 Oceans Act. Through this endeavour the federal government of Canada highlighted the Pacific North

Coast Integrated Management Area (PNCIMA) region of British Columbia as a priority conservation region. Figure 11.2 shows the limits of the PNCIMA.

Pacific Marine Analysis and Research Association (PACMARA)

The Pacific Marine Analysis and Research Association (PACMARA) has assisted the PNCIMA effort by bringing together scientists, researchers, representatives from all forms of government (federal, provincial, First Nations) and non-governmental organizations to develop and disseminate science-based information and knowledge for ecosystem-based marine management. PACMARA arose from a series of workshops mediated by the World Wildlife Fund during the period from 2003–2004. As of 2009, PACMARA has been seen as an unaffiliated and impartial institution, helping play an information broker role in the marine spatial planning and integrated management activities occurring in British Columbia (PACMARA website: www.pacmara.org).

One of PACMARA's most significant contributions to marine spatial planning to date has been the development and dissemination of a user-friendly handbook on strategic conservation using decision-support tools, particularly MARXAN (Ardron et al., 2008). As described in Chapter 3, MARXAN is one of several decision-support software tools available to the marine spatial planning community, and of all it is most in demand. The optimization software using simulating annealing algorithms to present heuristically a range of options for planning marine reserves, guided by the decision rules or design principles that the user stipulates. MARXAN can be used for undertaking zoning within an area as well, and one of its newer iterations, MARXAN with Zones or MARZone, is used specifically for this purpose. The MARXAN Good Practices Handbook clearly lays out the power of these decision-support tools, but also discusses their limitations.

While PACMARA has been providing a platform for information on data and tools relevant to marine planning, the non-governmental community based in British Columbia has been investing in joining together GeoBC and other government entities with NGOs in a spatial planning effort entitled BCMCA (British Columbia Marine Conservation Analysis). The BCMCA covers a swath of coastal and marine area on the northern portion of Vancouver Island, adjacent to the Priority Conservation Area encompassed by PNCIMA. The Living Oceans Society, based on the island of Vancouver, has been taking the lead in assembling databases on marine biodiversity, productivity, vulnerability and value, in order to steer the target area's management in a more sustainable direction. PACMARA has also assisted in this effort.

British Columbia Marine Conservation Analysis

The BCMCA is a collaborative project designed to provide information about marine biodiversity and human activity in BC's marine waters, with aims to create a comprehensive atlas and in-depth analysis of ocean values and uses to provide needed tools to help inform decisions about ocean management. Under the BCMCA project, ecological experts are being asked to answer these types of questions:

- What percentage of a species population needs to be represented in areas of high conservation value? (A range of values will be considered.)
- What is the minimum patch or population size to ensure population viability?
- How many patches are necessary to safeguard against localized impacts?
- What is the minimum distance between important patches identified in order that they provide high conservation value?

At the same time, experts on human uses are being asked to answer these types of questions:

- Does available data accurately and comprehensively capture the areas of interest or importance for the human use represented?
- How can the areas of importance to each human use be appropriately represented within MARXAN analyses?
- How can data on various human uses be appropriately synthesized and combined within MARXAN analyses?

In creating the comprehensive atlas of the BCMCA area, features such as broad ecological units, species, habitats and areas of ecological or human use identified through the ecological expert workshops held by the BCMCA or through discussions with the marine user community will be mapped. Priority sites will be identified using MARXAN decision support – these areas could well inform a zoning plan should that goal be decided upon by the interested parties.

According to its website (www.bcmca.net), the multi-organization BCMCA project team is assembling spatial data representing the distribution of ecological features and human uses in Canada's Pacific Ocean to inform integrated marine planning. These data are the foundation for the project's two products to date: an atlas of ecological and human use maps and analyses, using MARXAN and MARXAN with Zones, which identify areas of high conservation value that minimize overlap with areas important to human use. For each feature identified, experts have been asked to specify data sets that could inform the creation of a province-wide spatial representation of the feature; datasets are then assembled, with particular attention to sharing constraints. Often multiple data sets are combined to create a feature. Experts who recommended the feature and the data providers are then invited to review the spatial representation of the feature. Once suggested revisions are complete, the BCMCA will publish the feature as a map, as metadata and as a downloadable GIS data layer, wherever permissible through data sharing agreements. Regarding human-use mapping, a committee of user group representatives – called the Human Use Data Working Group – provides overarching direction and advice to the project team about the analysis of human use data (www.bcmca.net).

The atlas to be produced by the BCMCA will contain the following, according to the participants:

- Maps of all known and available ecological data layers, such as species distributions and habitats
- A richness map of ecological features (combined ecological data)
- Maps of all known and available human use data layers
- A richness map of human use components (combined human use data)
- A map showing both ecological and human use components
- Maps illustrating where data is currently lacking

To date, the BCMCA has collated and integrated more than 150 ecological datasets and 85 human-use datasets. Collating the best available spatial ecological and human use data is an enormous opportunity and challenge. The BCMCA project is adding value to largely disparate datasets by combining these data to create comprehensive maps, which can inform future marine planning efforts. The main challenge lies in sensibly combining datasets that were often collected using different methods and where metadata are not always available or complete (www.bcmca.net).

Other initiatives on Vancouver Island's west coast

On the west coast of Vancouver Island, another marine spatial planning effort has been taking shape, at a slightly smaller scale than the aforementioned initiatives. The West Coast Aquatics initiative was begun in 2008 to systematically look at the seven subregions of the island's western region and identify what sort of tools could be best applied in each. Such tools include not only ocean zoning but also 'objectives management' – something the West Coast Aquatics team considers a management approach that works by enhancing existing regulations instead of using ocean zoning. For two of the subregions (Clayoquot Sound and Barkley Sound), however, the West Coast Aquatics initiative has set out to work with local communities and other stakeholders to develop zoning plans.

Finally, a more focused initiative that is being led solely by the First Nations of British Columbia involves spatial management under something termed 'Tribal Parks'. These multiple-use parks capture the spiritual perspective that all the First Nations tribes seem to share, one which recognizes the connections between watershed and sea, and between man and all elements of nature. The latest of the Tribal Parks initiatives follows a long history of First Nations stewardship of lands. In 1984 the Tia-o-qui-aht proclaimed Meares Island as a Tribal Park to stave off logging interests, supported by their neighbours, the Ahousat First Nations. By late 1990, the Tia-o-qui-aht and the other Nuu-chah-nulth tribes of Clayoquot Sound had gained control of logging lands, and had entered into a memorandum of understanding with environmental groups, committing to preserving the pristine valleys of Clayoquot Sound. More recently, First Nations peoples have turned their attention to beyond the watersheds of their stewardship domain into the oceans – using the salmon as a symbol of the interconnectedness of ecosystems and the need to manage areas from watershed boundaries full out into the open sea.

In a related development, on 16 January 2010 the Haida Nation and Parks Canada jointly declared the world's first mountaintop to seafloor integrated protected area, committing the Council of the Haida Nation and the Government of Canada to share in the planning and management of waters in southern Haida Gwaii (otherwise known as Queen Charlotte Islands, British Columbia). This step extends protection provided by the Haida Heritage area 10km offshore, and paves the way towards the establishment of a National Marine Conservation Area Reserve. According to a Government of Canada press release, 'Haida traditional activities will continue and while conservation will be the focus, marine activities such as commercial fishing and recreational activities that meet conservation objectives will be permitted'.

All of these efforts, occurring at various scales and involving a multitude of agencies and NGOs, utilize marine spatial planning as a framework for making informed decisions about managing marine uses and access. Ocean zoning is a tool that has been used in most of the initiatives described, yet no standardized method for zoning has been agreed upon, and the use of spatial management has not occurred in a coordinated or strategic fashion. Should the government of the province of British Columbia, the governments of coastal First Nations, and the federal government of Canada decide to work together, a comprehensive ocean zoning plan could not only be one outcome – it could indeed be the stimulus to get all the vested interests to the table to work together.

References

Ardron, J.A., H. P. Possingham and C. J. Klein (eds) (2008) *Marxan Good Practices Handbook*, External review version, 17 May, 2008, Pacific Marine Analysis and Research Association, Vancouver, BC, Canada, 155pp. www.pacmara.org

British Columbia Marine Conservation Analysis (BCMCA) (2009) www.bcmca.net

Integrated Land Management Bureau (ILMB) (2007) *Marine Ecosections of British Columbia*, Spatial Analysis Branch, GeoBC, British Columbia, Canada

Living Oceans Society (2009) www.livingoceans.org

PACMARA (2009) www.pacmara.org

Watts, M. E., I. R. Ball, R. R. Stewart, C. J. Klein, K. Wilson, C. Steinback, R. Lourival, L. Kircher and H. P. Possingham (2009) 'Marxan with Zones: software for optimal conservation based land- and sea-use zoning', *Environmental Modelling and Software*, vol 24, pp1513–1521

12
Marine Spatial Planning in the US

T. Agardy

Marine conservationists meeting to plan protection for Palmyra Atoll, a US Territory in the Line Islands

Introduction

The first decade of the new millennium was an exciting time for marine spatial planning in the United States. The national government, reacting to both dire new warnings on the decline of its marine resources and the global ocean and to periodic calls to improve its governance of ocean areas, announced a new draft marine policy near the end of 2009. But long before that, some of the individual states had pioneered ocean zoning efforts in a push to improve marine management in state waters (to 3nm offshore). This was made possible by the 'new federalism' that emerged in the latter decades of the twentieth century, which ushered in a delicate power-sharing between the state governments and the national government of the United States. To date this power sharing has resulted in a series of unconnected, and as yet uncoordinated, initiatives, with each entity utilizing spatial planning only within their own jurisdictions and little thought given to the interface between adjacent (or sometimes overlapping) jurisdictions. This final case study chapter will begin with a look at two states on opposite coasts that have utilized ocean zoning in different ways and to different ends – but with the common end result of forcing the federal government's hand on marine spatial management.

Massachusetts Oceans Act

The waters of Massachusetts are fast approaching those in the North Sea in terms of the ever-increasing kinds of uses and pressures. Conservationists in the northeast US have called the quickening pace of ocean development in this part of the Gulf of Maine 'ocean sprawl', and have pushed to find and support tools that facilitate rational planning and minimize risks to the marine environment.

So it was that the 2008 decision by the state of Massachusetts to pass legislation creating the first comprehensive plan for state waters in the US was met with much support. The plan covers marine areas to 3nm, taking into consideration fisheries, renewable energy production and marine conservation interests, among many others. Legislators in Massachusetts have said that with the pace of proposed developments in state waters, including liquefied natural gas installations, offshore wind turbines and offshore aquaculture, they no longer had the luxury of being able to react to each proposal in isolation.

In 2007–2008, the Massachusetts Ocean Coalition travelled the coastal communities of Massachusetts to talk about participatory ecosystem-based management of the 1.6 million acres of the state's ocean environment. Nearly 5000 individuals expressed interest in improved marine management, paving the way for the passage of the Massachusetts Oceans Act in 2008 (MEAM, 2008). Among other measures, the Act stipulates that the state strategically plan for future development by considering comprehensive ocean zoning.

Rob Moir of the Massachusetts Ocean Coalition called the state effort 'revolutionizing environmental planning and management'. Every effort was made to listen to local communities and stakeholders, yet to plan ocean management with an ecosystem, wide-scale focus.

The 2008 Massachusetts Oceans Act states that: 'The Secretary, in consultation with the ocean advisory commission ... and the ocean science advisory council ... shall develop an integrated ocean management plan, which may include maps, illustrations and other media. The plan shall: (i) set forth the commonwealth's goals, siting priorities and standards for ensuring effective stewardship of its ocean waters held in trust for the benefit of the public'. The Act states an additional 14 principles or goals that the plan will target, including the last: '(xv) shall identify appropriate locations and performance standards for activities, uses and facilities allowed'. The common interpretation of this language is that ocean zoning will be used to address principle number 15.

The legislation mandated that a strategic plan be developed by the end of 2009, only a year-and-a-half after the Act passed. Setting such a short time frame was a gamble: imposing such deadlines on state planners could have set the intitiative up for failure, or even created a backlash among a public unused to rapid decision-making. However, it seems the compressed timeline has actually helped to make the planning process more efficient and participatory. Engaged stakeholders rose to the challenge, and financial resources and other forms of support came from those impressed with the ambitious goals of the state's planners. A draft plan was open for public comment mid-year in 2009; the final comprehensive plan for the state was made public on 5 January 2010.

The final plan establishes strictly protected areas in state waters, including eelgrass beds, submerged rocky reefs, whale migratory pathways and seabird areas. Ecologically sensitive areas constitute nearly two-thirds of the state's waters, where strong protections will be enacted. In addition, the plan allows up to 266 wind turbines in state waters: 166 in 2 designated commercial wind farm areas (or zones), and 100 scattered among smaller community projects. Community energy project planning will be under the remit of coastal planning authorities in 7 regions of the state's coast (www.mass.gov.eea).

Another outcome of the work on the Oceans Act was the formation of the Massachusetts Ocean Partnership, a non-governmental entity that grew out of the advisory commission to support the state in its strategic planning. Little movement towards comprehensive ocean zoning could have been achieved without such support.

In a sense the Massachusetts ocean planning process illustrates the adage of being united against a common enemy. There was little recognition of the inadequacies of marine management of ocean uses within state waters until two proposed developments in state waters (the building of a liquefied natural gas terminal outside Boston Harbor, and the proposed Cape Wind building of wind turbines in Nantucket Sound) made the state government realize that it had no plan and therefore no authority or rationale to oppose such developments. But in many ways the Commonwealth of Massachusetts was fortunate, having a liberal and environmentally aware political leadership, state government authorities willing to try things never tried before, a highly informed public used to active engagement in planning and a wealth of scientific information about the ecology, condition and effectiveness of existing management of the state's waters.

California Marine Life Protection Act

On the opposite coast, the conditions for allowing the development of new strategic planning for coastal waters also created an environment ripe for the application of marine spatial management and ocean zoning.

The California Marine Life Protection Act (MLPA) was passed in 1999, for the purpose of safeguarding the valuable and diverse marine ecosystems found within 3nm of the coast. The Marine Life Protection Act of 1999 directs the state to redesign California's system of marine protected areas (MPAs) to function as a network in order to: increase coherence and effectiveness in protecting the state's marine life and habitats, marine ecosystems and marine natural heritage, as well as to improve recreational, educational and study opportunities provided by marine ecosystems subject to minimal human disturbance (www.dfg.ca.gov/mlpa). There are six goals that guide the development of MPAs in the MLPA planning process:

1 Protect the natural diversity and abundance of marine life, and the structure, function and integrity of marine ecosystems
2 Help sustain, conserve and protect marine life populations, including those of economic value, and rebuild those that are depleted
3 Improve recreational, educational and study opportunities provided by marine ecosystems that are subject to minimal human disturbance, and to manage these uses in a manner consistent with protecting biodiversity
4 Protect marine natural heritage, including protection of representative and unique marine life habitats in Californian waters for their intrinsic values
5 Ensure California's MPAs have clearly defined objectives, effective management measures and adequate enforcement and are based on sound scientific guidelines
6 Ensure the state's MPAs are designed and managed, to the extent possible, as a network

To help achieve these goals, three types of MPA designations are used in the MLPA process: state marine reserves, state marine parks and state marine conservation areas (www.dfg.ca.gov/mlpa).

According to the California Fish and Game website, the MLPA directs the state to reevaluate and redesign California's system of marine protected areas (MPAs) to: increase coherence and effectiveness in protecting the state's marine life and habitats, marine ecosystems and marine natural heritage, as well as to improve recreational, educational and study opportunities provided by marine ecosystems subject to minimal human disturbance. The MLPA also requires that the best readily available science be used in the redesign process, as well as the advice and assistance of scientists, resource managers, experts, stakeholders and members of the public (www.dfg.ca.gov/mlpa).

California is taking a regional approach to redesigning MPAs along its 1770km coastline and has divided the state into five study regions (www.dfg.ca.gov/mlpa):

- North Coast: under way in 2010
- South Coast: 2008–2009 (regulations being reviewed)
- North Central Coast: 2007–2009 (regulations adopted 2009)
- Central Coast: 2004–2007 (regulations adopted 2007)
- San Francisco Bay: 2010–2011

Though not an ocean zoning initiative per se, the MLPA process has taken the first steps toward comprehensive ocean zoning, by defining ecoregions and then investigating what areas within those ecoregions most need protection or restoration. Should the state of California decide that ocean zoning of state waters is desirable, they will be poised to do so based on rigorous scientific information as well as a mechanism for participatory planning.

Federal government Ocean Policy Task Force

While the states of California and Massachusetts (and, more recently, Rhode Island) were forging ahead on marine spatial planning, the federal government seemed to be getting nervous. The former head of the White House Council of Environmental Quality under the previous administration, James Connaughton, had answered a question about ocean zoning from a member of the Federal Advisory Committee on Marine Protected Areas (MPA FAC) by saying he felt that ocean zoning was 'inevitable' (J. Connaughton, 2005). Even the new Obama administration has shied away from using the term 'zoning', committing instead to marine spatial planning (see below). It seems that during the years 2000–2008, the national government opted to sit back and watch the experiments of states like Massachusetts and California to see how they would deal with use conflicts and if ocean zoning gained traction.

The Obama administration established an Ocean Policy Task Force very soon after assuming leadership (likely at the urging of President Obama's choice for the head of the National Oceanic and Atmospheric Administration, Dr Jane Lubchenco). In part this was in recognition of the highly fragmented nature of US ocean policy: the nation's complex marine governance structure involves 20 federal agencies and approximately 140 ocean-related laws.

The task force developed a draft policy for public comment by mid-2009 (even more ambitiously than the Commonwealth of Massachusetts, the administration gave a scant six months for developing a draft policy. On 12 June, President Barack Obama established the Interagency Ocean Policy Task Force to develop recommendations for a national policy for US oceans, coasts and the Great Lakes, a framework for improved federal policy coordination and an implementation strategy to meet the objectives of a national ocean policy within 90 days). The interim policy, released in December 2009, embodies an entirely new approach to federal resource planning for oceans, coasts and the Great Lakes, outlining the process through which the federal government

Table 12.1 *Traditional, new, and expanding ocean, coastal and Great Lakes uses*

- Aquaculture (fish, shellfish and seaweed farming)
- Commerce and Transportation (cargo and cruise ships, tankers and ferries)
- Commercial Fishing
- Environmental/Conservation (marine sanctuaries, reserves, parks, wildlife refuges)
- Maritime Heritage and Archaeology
- Mining (e.g., sand and gravel)
- Oil and Gas Exploration and Development
- Ports and Harbours
- Recreational Fishing
- Renewable Energy (e.g., wind, wave, tidal, current and thermal)
- Other Recreation (e.g., boating, beach access, swimming, nature and whale watching, and diving)
- Scientific Research and Exploration
- Security, Emergency Response and Military Readiness Activities
- Tourism
- Traditional Hunting, Fishing and Gathering

will work with states, tribes, local governments and communities to decrease conflicts among competing users. It recommends the formation of a National Ocean Council to track the condition of the oceans and work with the White House to develop more effective marine management. According to a Council of Environmental Quality (CEQ) press release, the process is intended to improve planning and regulatory efficiencies, decrease their associated costs and delays, and preserve critical ecosystem function and services (CEQ, 2009).

The new policy thrust comes as the country faces ever-increasing reports of degradation, overexploitation, conflict and weakened natural systems unable to cope with large-scale changes coming on the horizon, such as those brought about by climate change. At the same time, access to and uses of oceans is increasing throughout the country, in both state and federal waters (see Table 12.1 on traditional, new and expanding uses in the coasts, oceans and Great Lakes of the US). The combination of increasing pressures and declining state of the ocean environment threatens the economy and well-being of the US and of all Americans.

One mechanism for reducing those use conflicts and making marine management more strategic is the establishment of nine regional planning bodies that will work together and with the administration in an integrated manner to develop short term and long range plans for managing ocean use (see Colour Plate 12.1). This recommendation mirrors that made by the MPA FAC in trying to facilitate the formation of a national system of MPAs. The thought is that regional bodies, comprised of federal and state government representatives as well as all relevant stakeholders, can better address management needs and utilize planning tools (such as ocean zoning) than centralized government. The Interim Framework for Effective Coastal and Marine Spatial Planning provides the conceptual framework for zoning, but shies away from the specifics on the zoning process (presumably this is somehting better left to the regional councils, in the CEQ's opinion). The document does, however, address federal authority for marine spatial planning, stating:

> *Federal statutes often include authorizing language that explicitly gives agencies the responsibility to plan and implement the objectives of the statutes. Moreover, several Federal statutes specifically authorize agency planning with respect to the ocean, coastal, and Great Lakes environments. Federal agencies and departments also administer a range of statutes and authorized programs that provide a legal basis to implement CMSP. These statutory and regulatory authorities may govern the process for making decisions (e.g., through Administrative Procedure Act rulemaking and adjudications) and not just the ultimate decisions made. The processes and decision-making CMSP envisions would be carried out consistent with and under the authority of these statutes. State, tribal, and local authorities also have a range of existing authorities to implement CMSP, although this will vary among and within regions. This framework for CMSP is to provide all agencies with agreed upon principles and goals to guide their actions under these authorities, and to develop mechanisms so that Federal, State, tribal, and local authorities, and regional governance structures can proactively and cooperatively work together to exercise their respective authorities.* (CEQ, 2009, p6)

Marine cadastre is a tool that is gaining popularity in US government planner circles, and it may well hold the key to ocean zoning potential. The US Multipurpose Marine Cadastre (MMC) is the Department of Interior's Minerals Management Service online tool for planning. The MMC helps coastal professionals locate the best available information for planning and mapping marine space and enables them to share that information quickly.

The task force policy on MSP also highlights some general principles for marine spatial planning that should faciliate ocean zoning initiatives undertaken at any level (Table 12.2).

It should be noted that this is not the first time the national government has tried to restructure itself to address marine management challenges. In the 1960s, the Stratton Commission evaluated how effectively fisheries, oil and gas, boating and shipping, and other uses were being managed in the country's territorial waters, and concluded that a complete governance restructuring was needed. The Stratton Commission recommended the formation of a Department of Oceans to centralize marine management and plan and regulate all uses under a single umbrella authority. Though this never came to pass, the idea of integrating management and using strategic tools such as ocean zoning is very central to the Obama administration's new interim policy. And though ocean zoning is deliberately not mentioned, the term 'marine spatial planning' appears 20 times in the framework document, and the steps in the marine

Table 12.2 *National Guiding Principles for Coastal and Marine Spatial Planning*

1	CMSP would use an ecosystem-based management approach that addresses cumulative effects to ensure the protection, integrity, maintenance, resilience and restoration of ocean, coastal and Great Lakes ecosystems, while promoting multiple sustainable uses.
2	Multiple existing uses (e.g., commercial fishing, recreational fishing and boating, marine transportation, sand and gravel mining, and oil and gas operations) and emerging uses (e.g., off-shore renewable energy and aquaculture) would be managed in a manner that reduces conflict, enhances compatibility among uses and with sustained ecosystem functions and services, and increases certainty and predictability for economic investments.
3	CMSP development and implementation would ensure frequent and transparent broad-based, inclusive engagement of partners, the public, and stakeholders, including with those most impacted (or potentially impacted) by the planning process and with underserved communities.
4	CMSP would take into account and build upon the existing marine spatial planning efforts at the regional, State, tribal, and local level.
5	CMS Plans and the standards and methods used to evaluate alternatives, tradeoffs, cumulative effects and sustainable uses in the planning process would be based on clearly stated objectives.
6	Development, implementation and evaluation of CMS Plans would be informed by the best available science-based information, including the natural and social sciences.
7	CMSP would be guided by the precautionary approach as defined in Principle 15 of the Rio Declaration, 'Where there are threats of serious or irreversible damage, lack of full scientific certainty shall not be used as a reason for postponing cost-effective measures to prevent environmental degradation.'
8	CMSP would be adaptive and flexible to accommodate changing environmental conditions and impacts, including those associated with global climate change, sea-level rise and ocean acidification, and new and emerging uses, advances in science and technology, and policy changes.
9	CMSP objectives and progress toward those objectives would be evaluated in a regular and systematic manner and adapted to ensure that the desired environmental, economic and societal outcomes are achieved.
10	The development of CMS Plans would be coordinated and compatible with homeland and national security interests, energy needs, foreign policy interests, emergency response and preparedness plans and frameworks, and other national strategies, including the flexibility to meet current and future needs.
11	CMS Plans would be implemented in accordance with customary international law, including as reflected in the 1982 Law of the Sea Convention, and with treaties and other international agreements to which the United States is a party.
12	CMS Plans would be implemented in accordance with applicable Federal and State laws, regulations, and Executive Orders. (CEQ 2009, pp7–8)

Source: CEQ, 2009

spatial process (Table 12.3) are essentially the same as those for developing ocean zoning plans.

The next decades will be a source of fascination for those interested in marine management and ocean policy. How this new US policy plays out, and the extent to which the public accepts and embraces comprehensive ocean zoning, remains to be seen. Nonetheless, all the pieces now seem to be in place in the US: an overarching federal policy that establishes regions as the geographic basis for marine spatial planning, field-tested methods developed

Table 12.3 *Essential Elements of the CMSP Process*

- Identify Regional Objectives
- Identify Existing Efforts that Should Help Shape the Plan throughout the Process
- Engage Stakeholders and the Public at Key Points throughout Process
- Consult Scientists and Technical and Other Experts
- Analyse Data, Uses, Services, and Impacts
- Develop and Evaluate Alternative Future Use Scenarios and Tradeoffs
- Prepare and Release a Draft CMS Plan with Supporting Environmental Impact Analysis Documentation for Public Comment
- Create a Final CMS Plan and Submit for NOC Review
- Implement, Monitor, Evaluate, and Modify (as needed) the NOC-certified CMS Plan

Source: CEQ, 2009

by some pioneering states and increasing recognition by the public of the need for rational and strategic planning to alleviate conflict and set marine resource use on a more sustainable pathway.

References

Council of Environmental Quality (CEQ) (2009) 'Interim Framework for Effective Coastal and Marine Spatial Planning', prepared by the Interagency Ocean Policy Task Force, Washington, DC, 35pp

Commonwealth of Massachusetts (2009) www.mass.gov.eea

Connaughton, J. (2005) Pers. comm., 'Remarks to the Marine Protected Areas Federal Advisory Committee', Thursday 17 February, Arlington, VA, USA

MEAM (2008) 'Comprehensive ocean zoning: Answering questions about this powerful tool for EBM', *Marine Ecosystems and Management*, vol 2, no 1, p2

State of California Fish and Game Department (2009) www.dfg.ca.gov/mlpa

13
Principles Underscoring Ocean Zoning Success

T. Agardy

Some would call this Fijian Island paradise

Ocean zoning commonalities and generalizations

Zoning has been called one of the simplest and most commonly used tools in coastal planning, as well as one of the most powerful (Kay and Alder, 2005). But while the concept of spatially separating and controlling different uses is accepted practice on land, comprehensive marine zoning is a nascent endeavour that is still largely experimental. Of course, much can be learned from smaller scale marine zoning that has been developed to segregate incompatible uses within marine protected areas. From these smaller scale examples, as well as some of the larger scale zoning initiatives described as case studies in Chapters 4–12, lessons can be derived and generic principles can be determined.

Comprehensive ocean zoning has great potential to allow both conservation of marine habitats and species as well as sustainable use of marine resources. As we have seen from the detailed examples provided in this book, and from experiences too much in their infancy to summarize in a case study yet illustrative of key principles nonetheless, zoning plans vary in complexity depending on the scale of the target area and the number of and kinds of uses that need to be accommodated. Often zoning is imposed in a top-down manner (nascent efforts in China and in Vietnam are examples), but in other cases zoning grows from the ground up, sparked by the interest of local communities in having a stronger hand in ocean planning and management. Most zones describe activities that are allowed, permitted and restricted or prohibited, but – and this is a key point – zoning regulations rarely state the levels of allowed or permitted activities. Rather, these limits are established on a site-by-site basis, if the issue of what constitute sustainable levels of use is addressed at all.

In general, the simpler the zoning scheme, the better understood by the public, and the better the compliance. Zoning boundaries must be clear, consistent and identifiable. Once zoning is in place, other coastal management tools can be linked to the zoning plan, such as seasonal restrictions, facility restrictions, permitting, licensing and quotas for resource extraction (Kay and Alder, 2005). Beyond these simple and somewhat common-sense generalities, the case studies described in Chapters 4–12 provide us with more specific lessons learned from translating ocean zoning from theory into practice.

Lessons derived from the case studies

From the oft-cited example of the zoning within the Great Barrier Reef Marine Park (GBRMP), we have learned that multiple use zoning that is specifically tied to multiple objectives is not only feasible, but easily understood by the public as well, even in a system as large and complex as the vast Great Barrier Reef (Chapter 4). Having the framework in place to do comprehensive, objective-oriented zoning – as provided by the Marine Park and its special authority – enabled the initial zoning and rezoning processes to occur. Although not legislatively mandated to periodically rezone, having legislation *allowing* rezoning (and having leadership amenable to it) created the enabling conditions to periodically update and improve management. This is the essence of adaptive management, which most would agree is one key to successful marine management.

We can also learn from mistakes made by practitioners struggling to put theory into practice on the ground. Returning to Australia's Great Barrier Reef, we see that by initially limiting the GBRMP Authority's sphere of management influence to only the marine portions of that ecosystem, managers were unable to adequately influence and mitigate the kinds of impacts arising from land use adjacent to the reef that contributed to its steady decline. Following the initial designation, the GBRMPA quickly saw that in order to be successful, better coordination between marine and land planners would be necessary – and they were able to forge an effective partnership between the national authority (the GBRMPA) and the state of Queensland. Arguably, this integration and coordination – had it occurred from the GBRMP's inception – could have staved off significant degradation of the reef ecosystems resulting from land-based sources of pollution, etc. What we learn from this is that comprehensive ocean zoning should be as *comprehensive* as possible, complementing and influencing both land-use and watershed management in adjacent areas.

In neighbouring New Zealand (Chapter 5), successes and failures with marine spatial planning have shaped the public debate about how ocean use should be regulated, which has in turn caused significant policy shifts at all levels of government. The negative reaction to ocean management and coastal access regulation among certain stakeholders – especially the indigenous Maori peoples – has constrained the New Zealand government in being able to comprehensively and strategically manage its ocean space and the resources over which it has jurisdiction. This emphasizes a general principle of marine planning: such planning should be as participatory as possible – right from the start. Better engagement of stakeholders, clearer communication about the problems that spatial management sought to address and a more transparent policy planning process could have facilitated an ocean zoning plan supported by the vast majority of stakeholders.

In the United Kingdom (Chapter 6), a variety of spatial planning efforts at different scales had neither amounted to a strategic approach, nor raised awareness among the public about the need for and utility of ocean zoning – until recently. The overarching Marine Bill, the Marine and Coastal Access Bill and the localized legislation being developed in Scotland and Wales are finally bringing the pieces together. Debate in Parliament has helped carry the message that a strategic approach to conserving valuable marine ecosystems and reducing conflicts among users is needed – and this has increased support for special protection measures and general marine planning. And piloting the methodologies for planning, as was done in the Irish Sea Pilot, drives a lesson home about the value of demonstration models and starting small in order to eventually think big.

Across the Channel in the rest of Northern Europe, the European Union and several coastal countries within it have established the mechanisms for putting maritime zoning into place (Chapter 7). In Belgium this zoning was predicated on an accepted practice of determining the relative value and sensitivity of different portions of the marine envionment, as well as an informative process to develop scenarios for spatial management in order to be able to predict its outcomes. But in many regards Belgium is a special case in northern Europe

– its EEZ is miniscule, and information about it abounds. For the nearby regional planning done by the OSPAR commission, the task of developing spatial management plans has not been nearly so easy, and the effort seems to have lingered in the default mode of identifying specially protected areas (a common first step in marine spatial planning). In contrast, the marine planning and ocean zoning efforts in Norway – also a member of OSPAR but not bound by the policies of the European Union – have been comprehensive and strategic, incorporating both conservation objectives and fisheries management needs into the management planning. But all of these myriad forms of using ocean zoning are still in their infancy; it may well be that the lesson that emerges from northern Europe is that time put into marine spatial planning – if it ends with plans and does not lead to implementation of zoning – may be time wasted.

Shifting to southern Europe and the much loved but overused Mediterranean Sea, we can learn from the small-scale zoning project undertaken at Italy's Asinara Island (Chapter 8). I chose this example of multi-objective zoning within a protected area for a case study primarily because I was involved in generating the zoning plans, but any number of other MPAs that were zoned scientifically and systematically could have provided similar lessons – the most basic of which is that optimizing or maximizing different objectives will lead to different patterns of zoning. Being able to articulate these specific objectives, develop a variety of zoning options to meet objectives in different ways and present the information needed to evaluate trade-offs and make informed choices paves the way to effective ocean zoning.

Further out to sea in the great, wide Mediterranean Basin, the RAC/SPA-led initiative to identify potentially important high seas sites (or sites in areas beyond national jurisdiction) could well lay the groundwork for comprehensive ocean zoning at some point in time (Chapter 9). Existing 'zones' in the Mediterranean Sea include marine protected areas, of course – but also extend to the GFCM-declared no-trawl zone at all depths below 1000m, and fisheries closed areas. The hierarchical analysis undertaken to identify ecoregion, Ecologically or Biologically Areas (EBSAs) within them, and priority areas within those provides a useful conceptual framework for other ocean zoning initiatives, especially those spanning jurisdictions across a wide region.

On the African continent, we have seen countries set the pace with strategic planning, legislation and governance arrangements to couple coastal to ocean management (Chapter 10). Here the conflicts between coastal uses (primarily subsistence use of resources and community demands on coastal space) and offshore uses (dominated by commercial fisheries and the energy industries) are intense and growing. Being proactive and using ocean zoning to further conservation, ensuring that use is as sustainable as possible and conflicts are kept to a minimum, is something that many African counties are using ocean zoning to do.

Being proactive is something that the government of Canada, the province of British Columbia, and the First Nations also ascribe to – and a spate of ocean zoning and marine spatial planning activities have occurred off the British Columbian coast as a result (Chapter 11). But it may well be that the enormous amount of energy expended on information-rich and scientifically driven marine spatial planning is at risk, given that an overarching framework for ocean zoning is lacking. A key lesson here is that frameworks do help integrate, and legal regimes tied to zoning can ensure that plans translate into action.

Finally, our review of some of the myriad ocean zoning initiatives around the world ends with news that is truly 'hot off the presses', at least at the time of this writing. In the United States, the leadership driving home marine spatial planning originated among the states (three in particular: Massachusetts, California and Rhode Island) – and the federal government has had to follow (Chapter 12). The Obama Administration has fully embraced marine spatial planning and is looking to devolve planning (presumably including ocean zoning) to regional councils, in which the time-tested state agencies can play a large role. This emphasizes an earlier point: that demonstration models do matter, but it also shows the importance of planning in concert across jurisdictional boundaries, whether those boundaries are state-to-national, nation-to-nation or territorial-to-high seas.

The European Commission's principles for marine spatial planning

The European Commission recognizes commonalities in the marine spatial planning process, as reflected in its ten principles for MSP, all of which equally apply to the ocean zoning process. These principles include: 1) Using MSP according to area and type of activity, 2) Defining objectives to guide MSP, 3) Developing MSP in a transparent manner, 4) Stakeholder participation, 5) Coordination to simplify decision processes, 6) Ensuring the legal effect of national MSP, 7) Cross-border cooperation and consultation, 8) Incorporating monitoring and evaluation in the planning process, 9) Achieving coherence between terrestrial and maritime spatial planning, and 10) A strong data and knowledge base (see Box 13.1 for further details). Yet while agreeing to common principles is laudable, even among a group as diverse in member states as the Eurpean Union, putting those principles into practice and agreeing on the means to do so is much more challenging.

When and how to use high-tech methods

One of the things that is most striking about how different entities have gone about harnessing the zoning tool for marine management has been the different degrees to which costly, high-tech methods have been utilized. For this reason, it is worth exploring whether there are commonalities to successful planning of zoning using certain methodologies.

New GIS and other spatial management tools, as well as decision-support tools such as MARXAN, have allowed planners to evaluate an almost infinite number of possibilities for zoning. When or multicriteria evaluation is coupled to GIS displays of environmental and use information, planners can see how to optimize for one set of objectives over another. All these and other

Box 13.1 The EC's key principles emerging from Maritime Spatial Planning Practice

1 Using MSP according to area and type of activity
Management of maritime spaces through MSP should be based on the type of planned or existing activities and their impact on the environment. A maritime spatial plan may not need to cover a whole area (e.g., EEZ of a member state).

For densely used or particularly vulnerable areas, a more prescriptive maritime spatial plan might be needed, whereas general management principles might suffice for areas with lower density of use. The decision to opt for a stricter or more flexible approach should be subject to an evaluation process.

MSP operates within three dimensions, addressing activities (a) on the seabed, (b) in the water column and (c) on the surface. This allows the same space to be used for different purposes. Time should also be taken into account as a fourth dimension, as the compatibility of uses and the 'management need' of a particular maritime region might vary over time.

2 Defining objectives to guide MSP
MSP should be used to manage ongoing activities and guide future development in a sea area. A strategic plan for the overall management of a given sea area should include detailed objectives. These objectives should allow arbitration in the case of conflicting sectoral interests.

3 Developing MSP in a transparent manner
Transparency is needed for all documents and procedures related to MSP. Its different steps need to be easily understandable to the general public. This will allow full information to all parties concerned and therefore improve predictability and increase acceptance.

4 Stakeholder participation
In order to achieve broad acceptance, ownership and support for implementation, it is equally important to involve all stakeholders, including coastal regions, at the earliest possible stage in the planning process. Stakeholder participation is also a source of knowledge that can significantly raise the quality of MSP.

5 Coordination within member states – simplifying decision processes
MSP simplifies decision-making and speeds up licensing and permit procedures, for the benefit of maritime users and maritime investment alike. Coordinated and crosscutting plans need a single or streamlined application process and cumulative effects should be taken into account. The internal coordination of maritime affairs within member states proposed in the Guidelines for an Integrated Approach to Maritime Policy should also benefit the implementation of MSP.

Developments in the member states (e.g., UK and Scottish Marine Bill) demonstrate that national authorities are keen to reap these benefits through the establishment of a coordinating administrative body.

6 Ensuring the legal effect of national MSP
MSP does not replicate terrestrial planning at sea, given its tri-dimensionality and the fact that the same sea area can host several uses provided they are compatible. However, in the same way that terrestrial planning set up a legally binding framework for the management of land, MSP should be legally binding if it is to be effective. This might also raise the issue of the appropriate administrative framework for MSP.

> **7 Cross-border cooperation and consultation**
> Cooperation across borders is necessary to ensure coherence of plans across ecosystems. It will lead to the development of common standards and processes and raise the overall quality of MSP. Some organizations such as HELCOM have already started this work.
>
> **8 Incorporating monitoring and evaluation in the planning process**
> MSP operates in an environment exposed to constant change. It is based on data and information likely to vary over time. The planning process must be flexible enough to react to such changes and allow plans to be revised in due course. To meet these two requirements, a transparent regular monitoring and evaluation mechanism should be part of MSP.
>
> **9 Achieving coherence between terrestrial and maritime spatial planning – relation with ICZM**
> Achieving consistency between terrestrial planning (including coastal zones) and maritime planning systems is a challenge. Coastal zones are the 'hinge' between maritime and terrestrial development. Drainage areas or land-based impacts from activities such as agriculture and urban growth are relevant in the context of MSP.
>
> This is why terrestrial spatial planning should be coordinated with MSP. The respective services should cooperate and involve stakeholders so as to ensure coherence.
>
> **10 A strong data and knowledge base**
> MSP has to be based on sound information and scientific knowledge. Planning needs to evolve with knowledge (adaptive management). The Commission has started several scientific and data gathering tools that will assist MSP in this process. These include a European Marine Observation and Data Network (EMODNET), an integrated database for maritime socioeconomic statistics (currently under development by ESTAT), the European Atlas of the Seas (to be delivered in 2009) and the Global Monitoring for Environment and Security (Kopernikus).
>
> *Source:* Adapted from Verreet, 2009

tools, however, require time and resources to pick suitable algorithms, gather information to feed into the analysis, crunch the data and evaluate results. Several coastal planners (this author included) have questioned whether these high-tech methods yield better results than more Delphic or expert opinion-driven procedures, in which experts on both environment/ecology and socioeconomics are asked to brainstorm to come up with a zoning plan. While this question remains unanswered, it is important to consider the possibility, and also to take note of the widespread public suspicion that greets those planners who use such complicated tools to develop their plans that the public is mystified by the process.

There are, however, other issues with the use of high-tech methods, beyond problems of perception. The assumptions that one uses to make choices must be sound and clearly articulated, to be sure that decision rules are understood to be based not only on the best possible science, but also scientific consensus. This did not seem to be the case with the operating principles used by the Great Barrier Reef Marine Park Authority, some of which were based on questionable assumptions. For instance, the first biophysical operational principle states that no-take zones should have a minimum size of 20km along the smallest dimension. While this may indeed be an important criterion for the purposes of achieving public understanding and streamlining enforcement, the size constraint is hardly an agreed upon biophysical principle and is dubious when applied across a lot of highly different types of ecological communities. And with recent findings concerning larval dispersal and retention in reef systems, it may emerge that many smaller reserve areas are preferable to fewer big ones.

Another issue that always pertains to standardizing data for use in GIS and decision-support software is recognizing the extent to which inconsistencies exist in data quality. For while some nearshore communities like coral reefs and kelp forests are well understood and can be characterized ecologically and accurately valued for providing various services, most other marine community types are understudied. Geographic information systems and high-tech decision-support tools may gloss over these discrepancies, and indeed obscure the need for practising adaptive management in less well studied areas in order to gain better ecological information over time.

Additional principles

The following general recommendations are distilled from reviewing the zoning processes described in this book, as well as other marine spatial planning in various parts of the world. Like the EC's marine spatial planning principles outlined above, these recommendations are meant to help guide ocean zoning efforts in all their diversity of scales and geographies. However, the specific process that is utilized to develop and implement a zoning plan can and should vary from place to place.

1. **The overarching goal of zoning** – whether it be better integrating management over a wide area, increasing protection for biodiversity, maximizing tourism or recreational values, etc. – must be clearly articulated from the start. This goal-setting, as is the case for planning in general, should be as participatory as possible
2. **The choice of tools** to be utilized and the process followed should be a function of the geographic scale of the area to be zoned, the time allotted (or expected) in which the zoning process is to occur and the availability of scientific or quantitative information
3. Given the generally short time frames allotted for planning processes and the ecological uncertainties or inconsistencies, it may make sense to **combine various approaches.** In very large geographic scales, one might utilize a Delphic approach or multicriteria evaluation to highlight areas where more data-driven and objective analysis might be needed
4. **Explaining the use of tools** is important, particularly if high-tech decision-support tools are used, and significant effort should be devoted to explaining the tools to the public early in the process
5. **Assumptions behind designation** of operating principles and criteria for zoning should be clearly stated and, if possible, reflect scientific consensus

6 **Uncertainties should be highlighted,** not obscured, and should provide the basis for future adaptive management-related and basic research protocols
7 Environmental or ecological information and determinations of values of areas/resources should be gained not only from scientific study but also from **user knowledge**
8 **Weigh simplicity against effectiveness:** while it is true that the simpler the zoning plan the better understood, there is a fine line between aiming for understanding and compliance and creating an ineffective management system
9 **Do not underestimate the ability of users** to understand more complicated zoning plans and know where they are in the sea, despite the absence of fences and posted signs that make zoning on land so much easier
10 Ocean zoning should, to the maximum extent practicable, **link to land-use zoning**
11 As marine zoning becomes accepted practice, planners should look at ways to **couple the time dimension to the spatial one,** that is, to incorporate seasonal or dynamic zoning where it may be needed
12 **Zoning plans should never be considered cast in stone,** but rather dynamic entities that need revisiting periodically, both as environmental conditions change and as new information is obtained through basic research and adaptive management

That there are so many commonalities in the approaches that government agencies, communities and non-governmental organziations and academia have taken in launching ocean zoning processes suggests these many principles are indeed central to the success of the zoning effort. Because zoning should be considered more a process than a product – or, in common parlance, 'it's the journey that matters, not the destination' – it is interesting to note how very similar the journeys have been, whether in using zoning to design small-scale protected areas or to comprehensively zone entire EEZs to accommodate all possible uses. The most challenging aspect of those journeys has not been in the planning phases, but in the implementation – the subject of the next and penultimate chapter.

References

Kay, R. and J. Alder (2005, 2nd edition) *Coastal Planning and Management*, Taylor and Francis, Abingdon, UK, and New York

Verreet, G. (2009) 'Ecosystem based approach, MSP and link with the Marine Strategy', available at eurlex.europa.eu/LexUriServ/LexUriServ.do?uri=CELEX: DKEY=483715:EN:NOT

14
Implementing Ocean Zoning

T. Agardy

Patmos Island landscape, Greece

From plans and maps to regulating use at sea

At its most basic level, zoning involves drawing lines on a map and developing a set of use regulations pertaining to each delimited area (Hendrick, 2005). How to draw those lines, and what uses to regulate within them, remain the challenges for planners. But how to then implement spatially explicit regulations in each zone in a way that promotes compliance and support for management, and the ability and willingness to adapt as needed, remains the challenge for management agencies.

Implementation of ocean zoning plans has two aspects: setting up good governance to oversee the process of turning plans into action and operationalizing the management itself. The latter includes establishing and updating regulations; creating and distributing zoning maps; communicating with user groups and the general public; installing signage, buoys, etc. (where necessary); patrolling or undertaking other forms of monitoring and surveillance; setting up and maintaining a system for enforcement (and publicizing not only the rules and regulations but the fines associated with transgressions); and carrying out research to monitor the efficacy of management, as well as better understand the ecology and dynamics of the ecosystem.

Implementing a zoning plan does not require a wholesale restructuring of government, but it does require that attention be paid to the governance structures in place and the capacity of government to work with, in addition to regulating, user groups. The integration of management (and planning) that is so needed, and which ocean zoning can deliver, is rarely achievable with conventional government structures, at least those typically found in coastal countries around the world. Rather, effective governance requires greater coordination within government and improved governance beyond government.

There are instances in which marine spatial planning leads to ocean zoning plans that are robust in being science-based and strategic in theory, yet fail in implementation. The reasons for such failures are myriad (such as when common ocean zoning principles are ignored – see Chapter 13), but often a key factor that makes or breaks a zoning plan is whether the appropriate governance regime is in place to execute it. A zoning plan that exists only in concept or on paper is akin to a 'paper park' that lacks active governance by the responsible authority, fails to influence the public's perception that the uses of the area need to be regulated and falls short of managing access to and use of resources in the area. I begin this chapter with a discussion of what constitutes good governance and then turn to the day-to-day management needed to make zoning plans a reality.

Governance issues

The word governance is most commonly associated with government. However, modern-day political science considers governance as ocurring not only in the realm of government but also in two other institutions: civil society and

Table 14.1 *Major expressions of governance*

Government
- laws and regulations
- taxation and spending policies
- education and outreach

Marketplace
- profit-seeking
- ecosystem service evaluation
- eco-labelling and green products

Institutions and organizations of civil society
- socialization processes
- constituency building
- co-management

Source: Juda and Hennessey, 2001

markets. These forms of governance interact with one another through the mechanisms listed in Table 14.1.

The National Research Council's Committee on International Capacity Building for the Protection and Sustainable Utilization of Oceans and Coasts has defined the term governance to encompass the values, policies, laws and institutions by which a set of issues are addressed (NRC, 2008). Governance addresses the formal and informal arrangements, institutions and mores that structure and influence:

- How resources or an environment are utilized
- How problems and opportunities are evaluated and analysed
- What behaviour is deemed acceptable or forbidden
- What rules and sanctions are applied to affect how access to natural resources is allocated and how resources and space are utilized

The process by which science-based knowledge is incorporated into planning and decision-making processes, and the way the sometimes conflicting interests and values of the people and institutions affected are negotiated, is best described as a *governance process* – whether this process occurs within government or outside it (NRC, 2008).

Governance by government

Zoned ocean areas within a region can be administered by a variety of means – by a single overseeing state or federal agency that designs the zoning plans, by a coordinating body that ties together areas variously implemented by different government agencies, by an international body with representatives of government from all the countries sharing regional waters or by an umbrella framework such as the Biosphere Reserve Programme (UNESCO, 1996). The latter has benefits in that local communities become a part of the network, ecologically critical areas are afforded strict protection while less important or less sensitive areas are managed for sustainable use and the biosphere reserve designation itself carries international prestige (and can be used to leverage funds) (Agardy, 1997). However, such United Nations-affiliated programmes do not uniformly command respect among nations of the world, so it is difficult to see how such programmes could provide adequate governance for zoned ocean areas.

For shared coastal and marine resources, regional agreements may prove more effective, especially when such agreements capitalize on better understandings of costs and benefits accruing from shared responsibilities in conserving the marine environment (Kimball, 2001). One such example – a regional body overseeing human impacts on the terrestrial/freshwater environment in a large region – is the Mekong River Commission (see www.mrcmekong.org); this is a potential model for marine environments as well.

Municipal and other local governments are strengthening in power and in their ability to influence environmental and public policy at broader levels of government as well (note how the state of California, for instance, has affected US national energy and environmental policy, providing leadership through example). Divesting power and authority to the local level in an effort to find ways to co-manage is not something most governments are in a rush to do but are doing nonetheless as the burdens of management and regulation become overwhelming.

In the United States and Canada, and in some other countries as well, individual states or provinces have taken the lead on exploring the concept of zoning. In such situations, it is likely that zoning plans will first be implemented at the state level within the jurisdictions that the states control (e.g., 3nm in the US). Whether ocean zoning in these locations will be housed under the state coastal planning and management agencies (Office of Coastal Zone Management, Department of Environment or similar), or whether new ocean zoning entities will need to be established, will depend on the capacity of existing agencies and the resources available to the state. In a recent development (7 January 2009), the governors of six eastern US states appealed to the federal government for funding to carry out the activities called for in the newly announced national ocean policy.

With the increasing attention and interest in ocean zoning, it is likely that the federal government will follow the lead of the states and assess what kind of zoning should occur where in federal waters. The George W. Bush administration had already acknowledged that zoning is a part of the future for US waters, although not publicly (Connaughton, 2006), while the Obama administration recently made a big push for marine spatial planning, forming an overarching National Ocean Council and eight regional working councils to investigate the possibility of using ocean zoning, among other tools.

Governance by civil society in partnership with government

There is great potential in fostering ocean management, decision-making and leadership at the local level through ocean zoning initiatives. Coastal communities have ever-greater reliance on the coasts and oceans. This reliance can be transformed into stewardship if users are given the opportunity, since a key component of responsible behaviour is the recognition of how such behaviour is to one's own benefit. Stewardship means participation in decision-making and management, but it also means taking on the responsibility for care. This, in my opinion, is best achieved through local involvement in all forms of governance, including not just 'government' but also civil society and the private sector.

The interesting and new development in ocean management is thus not increasing stakeholder involvement, per se, but rather increasing *local* stakeholder involvement. Rather than being the recipients of plans for management of local coastal and marine areas, local communities and institutions have become active participants in goal-setting, planning, management, monitoring, enforcement and evaluating outcomes (Agardy, 2008a). Ocean zoning schemes can provide a mechanism for enhancing this aspect of governance.

The emerging sea change in participatory management down to the local level (or perhaps 'back down to the local level' would be more correct) is exemplified in ways too extensive to treat comprehensively here. But a few examples can point to how differently this local involvement manifests itself. For instance, there is the growing movement of communities hiring watchdogs to monitor compliance with existing pollution and/or fishing regulations and to publicly blow the whistle when infractions occur, such as occurs with the pooling of community funds to underwrite the salary and expenses of a 'Soundkeeper' in Long Island Sound (US). Local non-governmental organizations and even 'unorganized' community institutions are taking on the stewardship of marine protected areas all over the world, from Mafia Island in Tanzania and Apo Island in the Philippines to Abrolhos Marine National Park in Brazil. Local community groups have begun to foment the development of voluntary coastal and marine zoning plans, such as is occurring in the San Juan Islands of northwestern United States. Community development banks, providing loans to underinvested communities in order to reinvigorate local economies and create healthier local environments, are springing up in developing and developed countries alike (Agardy, 2008a).

Localized decision-making and co-management can pave the way to better governance, and clearly better and more diversfied governance of coastal and ocean areas is needed. Local involvement allows for better governance beyond government, such as creating and sustaining truly effective, goal-oriented and streamlined non-governmental institutions (Agardy, 2008a). And the engagement of the private sector, which will be discussed in greater detail in the following section, can also be strongly catalysed at the local level, through Chambers of Commerce, trade associations and individual companies with vested interests in sustaining the local environment.

Other stakeholders beyond local communities can be brought into the governance process – and in general, the wider and more diverse the stakeholder representation, the less the chance that the management of an area will be constained by those opposing regulations (though the time needed for planning, managing and making amendments to management may necessarily be longer than in those situations where only a few stakeholder groups participate).

As in the planning process, appropriate stakeholders include both affecting and affected parties (that is, those whose activities, whether on land or at sea, affect ecosystems, and those whose livelihoods or well-being are affected as a result). Such stakeholders can be thought of as being in three general categories: 1) the local community, including civil, non-governmental and labour organizations; 2) the public sector, including central, provincial/state and local government, public service agencies and publicly chartered institutions; and 3) the private sector, including fisheries, aquaculture, energy production and manufacturing industries, waste disposal, tourism, agriculture and forestry.

Each of these entities can be brought into a formalized co-management process, which is defined as a collaborative arrangement in which a community of local resource users, local and senior governments, and other stakeholders share responsibility and authority for managing a specified natural resource.

The willingness of communities and stakeholders to engage in co-management to implement ocean zoning schemes depends, at least partially, on their abilities to envision something worthwhile coming from their efforts, as well as the perceived costs of those efforts (S. Farber in NRC, 2008). Sustained good governance requires a continual reinforcement that efforts and costs are worth it. Issues likely to be of concern to people are the economic, social and ecological returns from capacity building, along with the costs associated with the effort of building and sustaining community interests, monitoring and enforcing commitments, educational requirements and meeting institutional needs. While many of the costs and benefits are narrowly economic, others may be more social, such as equity issues associated with access to managed ecosystems.

For instance, co-management arrangements in community-based MPAs in Puget Sound, Washington (US), allow the state government to save costs by having the community patrol the sites and record infractions. Similar co-management allows effective management of the Banco Chinchorro Reserve, far off the coast of Mexico's Yucatan Peninsula (NRC, 2008). The use of volunteers to both patrol and help undertake monitoring to gauge changes in environmental conditions (and to see if management objectives are being met) can significantly reduce the costs of MPA management and its demands on human resources. However, such volunteers need careful management to ensure that any monitoring is effective and that they do not take on an unwarranted enforcement role.

The role of markets and the private sector

With the exception of the highly regulated energy industry, the private sector has to date avoided becoming involved in management of the oceans. That is, while industries such as commercial fisheries and shipping have been granted use privileges, there has been little private-sector interest in investing in coastal and ocean management measures to sustain the benefits that well-managed ecosystems could provide. This is true for myriad reasons: 1) there is an enduring assumption that the public sector will manage ocean resources and space sufficiently, 2) without the development of markets, the vast majority of social and economic values associated with coastal and ocean areas remain unaccounted for in capital market transactions, and 3) the very nature of oceans with its commons property regime has required innovative new approaches to involving the private sector that are untested at large scales (Agardy, 2008a).

Surveys of how coastal ecosystems are currently protected show that innovative financing mechanisms that tap into the private sector are very few and far between. With a few exceptions, the protection of these areas has generally fallen to the public sector. Government agencies regulate coastal land use, freshwater and wetlands use, maritime activities, resource extraction, and the protection of threatened species and critical habitat. Yet the funds available to manage the coastal and ocean areas, both terrestrial and marine, are generally inadequate. There are inequities in the way coastal areas are managed, such that taxpayers shoulder the costs of protection, while many industries receive an almost free ride in taking advantage of the benefits coastal ecosystems provide. This indicates that there is much room for a dramatic shift in the engagement of business in ocean management in general and the implementation of ocean zoning in particular.

With growing recognition of the value of ecosystem services – the things nature does for us for free – new sources of potential market demand are emerging for ecosystem services. Payments for ecosystem services markets are expanding rapidly on land, and there is growing awareness of market potential for carrying this seaward among business, investment and conservation communities (Agardy, 2008b). The business community is beginning to see real opportunities for moving from single transactions to the full-blown development of markets. Some of these markets will likely be modelled on the carbon emissions markets of the EU and the Chicago Exchange, with polluters buying credits from companies or property owners that have taken steps to reduce water-quality impacts. Other markets which are likely to develop include wetlands mitigation-type markets extended into the marine zone, marine biodiversity offsets, marine species banking and habitat management supported by the very industries that realize the benefits: the fishing industry protecting nursery habitats, for instance, or the tourism industry protecting mangrove for its water filtration and buffering roles. Ocean zoning potentially has a huge role to play in the development of such markets, for two important reasons: 1) zoning establishes clear rights and repsonsibilities, which is a reassurance to business investors, and 2) zoning plans could include 'trading zones' where payments for ecosystem services transactions could be established.

New examples are emerging of public/private partnerships, as governance arrangements become more creative and effective. In one example, Gaines and Costello (2009) discuss the relationship between marine protected areas (MPAs) and territorial user right fisheries (TURFs). Though seemingly contradictory, both approaches are commonly advocated as solutions to failures in fisheries. Protected areas (or zones of protection) limit harvest to certain areas, but may enhance profits outside via spillover. TURFs incentivize local stewardship, but may be compromised when the TURF is too small to retain the offspring of adult fish in the TURF. If one assumes strategically sited MPAs may be an effective complement to spatial property-rights based fisheries, increasing both fishery profits and abundance, then it is logical that codifying these public/private partnerships within a zoning scheme could be maximally strategic and effective.

Is special legislation needed to implement zoning?

Many countries have targeted legislation that allows the creation of special zones such as marine parks, sanctuaries, reserves or other sorts of marine protected areas. Sometimes this legislation is developed systematically and strategically, laying the groundwork for some sort of national system of MPAs. Other times, MPA legislation arises because plans for designating a single MPA are in the works, but the MPA cannot be implemented without new legislation. An example of this is the case of the Mafia Island Marine Park in Tanzania, the designation of which required a special Act of Parliament (Agardy, 1997).

At another scale, China has developed an entire marine policy around ocean zoning, which came into effect with the 2002 passage of the Law on the Management of Sea Use (adapted from Li, 2006, cited by Ehler and Douvere, 2007). In developing marine spatial plans under this law, China has adopted three key prinicples: 1) the right to the sea-use authorization system, 2) a marine functional zoning system, and 3) a user-fee system. Through a classic top-down scenario, the government of China has tasked the State Oceanic Administration with determining zoning plans for the nation's waters, parcelling the territorial sea and Exclusive Economic Zone (EEZ) into four functional zones. Since China considers the ocean a state-owned asset (Ehler and Douvere, 2007), anyone benefiting economically from use of the sea must pay a fee, and anyone using ocean areas or resources must abide by the zoning rules. However, like the new federalism that arose to allow power-sharing between the federal (national) government and the states in the US, China's Law on the Management of Sea Use allows a two-tiered management system in which applications for ocean use in permitted areas have to be approved by the provincial government as well as national authorities.

There are cases, however, where the designation of protected zones does not require special national-level legislation. When the Great Barrier Reef Marine Park was established in Australia, the federal government created an Act that would allow the park to be implemented, along with a management authority known as the Great Barrier Reef Marine Park Authority (GBRMPA). The GBRMPA works closely with state agencies and the private sector in administering park regulations and in developing and adjusting zoning plans (see Chapter 4 on the Great Barrier Reef Marine Park). Furthermore, there are numerous other Federal and State Acts and Regulations which enhance the protection and conservation of the Great Barrier Reef, including legislation covering such aspects as a prohibition on mining, shipping, pollution, navigation aids, native title, aquaculture, historic shipwrecks, sea installations and heritage protection (Day, 2009).

Tenure of marine areas and some forms of traditional use can also be effective coastal conservation interventions, even when these patterns of sustainable use of marine and coastal resources occur outside of conventional protected areas (Curran and Agardy, 2002). Common property and common property management regimes have evolved in many coastal communities, and in some cases have been shown to be much more effective in keeping resource use to sustainable limits than have conventional, top-down methods (Curran and Agardy, 2002; MEA, 2005). Legitimizing such traditional uses remains an issue in many coastal countries, and recently non-governmental organizations have begun to liaise with governments to help codify use rights for local communities. Ocean zoning can help in this regard – through the creation of traditional use zones, for instance, or by building a multiple use zoning plan on the traditional management regimes and authority structures of local communities.

Management: operationalizing ocean zoning

Best practices in marine management are those that adequately fit the circumstances of the place, including environmental and ecological circumstances, the cultural and sociopolitical context, and the economic and logistic feasibility of undertaking management. Given that these circumstances vary so widely, it is safe to say there is no single approach to management of zones in a zoning scheme that can be universally applied, just as there is no single MPA management scheme that is universally applicable. Instead, there are common principles that emerge from well-designed and successfully implemented reserves (see Chapter 13).

First is the question of whether the zoning plan has clearly articulated objectives, which the public understand and accept. A good understanding of threats is required; regulations in each zone must address current and/or prospective threats.

Management must not only reflect the desired objectives and rise to the challenge of meeting threats head-on, but also the logistical realities in protecting marine areas offshore from both direct and indirect human impacts. Surveillance costs can be exorbitantly high, and enforcement is made difficult not only by surveillance problems but also by problems created by a sometimes uninformed public that does not recognize the invisible boundaries of a protected area or understand the rules. Bringing users into the management process can help alleviate some these issues and economize on costs. This is one form of what is referred to as co-management.

Another element common to all successful marine spatial management has to do with communications and education. When user groups are unfamiliar with the process of how a zoning plan was designed, do not know its objectives or indeed do not even know the zonation exists, there is little chance of having lasting support or gaining compliance. For strictly protected zones analogous to marine reserves, it may make sense to have certain user groups do some kind of training on regulations and the reasons for them (e.g., as is the case for Bonaire Marine Park in the Netherlands Antilles; Agardy, 1997).

Implementation of designated ocean zoning plans must be followed by what is one of the most crucial keys to MPA success: monitoring and evaluation. Monitoring can determine how the site is faring in light of deteriorating environmental conditions overall, and it can provide specifics on ecological changes happening inside the zone. But an extremely important purpose for monitoring is to gauge whether management objectives are being met. In this regard, a monitoring and evaluation protocol should be developed during the planning phases of the ocean zoning process, and put into motion as soon as the zoning is implemented. Recommended Further Reading lists several key publications on ocean zoning, including guidelines on monitoring and evaluation in the marine environment. Guidance contained in these and other publications points to the necessity of tracking not only the changes in environmental condition and ecology that zoning measures bring, but also the impact that zoning plans have on society, including economics, institutions and perceptions (for further reading, see Chapter 2).

There is no room in the scope of this book to review all effective types of marine management that might be put into practice through an ocean zoning scheme. However, there are generalizations about how management should be integrated through ocean zoning, including the need to coordinate sea-use management with land-use management, strictly protecting the most ecologically critical or vulnerable habitats.

Integrate land and sea-use management

Integrated management of watersheds, land-use planning and impact assessment are keys to protecting coastal sites. Complex problems require comprehensive solutions. For this reason, tackling the issues of loss and degradation of marine areas by addressing single threats to these environments will not be productive. Effective management of these crucial areas means coordinated pollution controls, development restrictions, fisheries management and scientific research. To be truly holistic, integrated management of marine nursery areas also requires complementary watershed management and land-use planning to ensure that negative impacts do not reach nursery areas from afar.

To fully understand and quantify the trade-offs to be made when coastal development, environmental degradation through waste discharge, or exploitation of marine areas occurs, environmental impact studies should take into account the full value of these ecologically critical areas. Zoning plans and permitting procedures for development that is potentially environmentally harmful should take into account the costs of losing the ecosystem processes and services that these areas provide (Villa et al., 2001).

Create fully protected MPAs in ecologically critical areas

When marine areas have been identified as particularly important, necessary steps should be taken to conserve these habitats and the species within them through strictly protected zones within the zoning scheme. Such protected areas may be small fisheries reserves in which resource extraction is prohibited, or they may occur in the context of larger multiple use areas. Marine protected areas networks are vital in this regard, acting to safeguard the most critical areas of the marine environment such as marine nursery areas throughout a geographically large area (Hyrenbach et al., 2001).

In order for marine protected areas to succeed in meeting the objectives of conserving habitats and protecting fisheries and biodiversity, management of these areas should address, if possible, all the proximate threats to marine and coastal areas. In most coastal habitats around the world, the threats are multiple and cumulative over time. Thus, protected areas that address only one of these threats in a piecemeal fashion will usually fail to conserve the ecosystem or habitats and the services they provide. Though complicated, planners should look not only at the specifics surrounding the threats affecting an area, but also the bigger picture (Agardy, 2005).

Manage fisheries on an ecosystem basis

Historically, most marine fisheries were managed on a stock-by-stock or fishery-by-fishery basis. Increasingly, people have begun to realize this single-species approach is inefficient in conserving the complex ecological processes in marine and coastal systems. Among fisheries management agencies and conservationists alike, a new push is now on for ecosystem-based fisheries management – management that looks at multi-species interactions and the entire chain of habitats these linked organisms need in order to survive and reproduce. The protection of nursery habitat must figure very prominently in ecosystem-based fisheries management (NRC, 1999). As a result of this rather new approach, conservationists and spatial planners have begun to work with fisheries' biologists and managers – spanning a gap between disciplines. In a sense, zoning that stipulates kinds and levels of permitted fisheries activity can be thought of as 'performing zoning', in which it is not the activity that is regulated per se, but the impact of the activity (Young, 2006).

Control or mitigate pollution

Pollution of coastal areas is a very significant cause of loss of important ecosystem services in much of the world. The overfertilization of nearshore waters via land-based sources of pollutants and via waste disposal into rivers and coastal zones is a particularly acute problem. One method of mitigation is to conserve, reconstruct or construct wetlands that act as filters of these pollutants before such compounds enter the coastal environment. Another is to encourage land-use practices such as buffer strips in agriculture and forestry to prevent the run-off of fertilizers, sediments, etc. As for hydrocarbons and other toxins, the way that municipal waste and storm run-off are treated should be improved. Finally, dredging operations should be assessed for the degree to which they may release pollutants into the water column. All of these impacts should be considered in ocean zoning schemes (the extent to which mitigation should be an element of management depends on the degree of threat posed by pollution in the place to be managed using ocean zoning; this will be revealed through the sorts of threats analysis described in Chapter 3).

Restore key areas

Some marine habitats such as mangrove forests and marshes can, in theory, be restored once destroyed. Such restoration is risky, however, since it has yet to be shown that the full range of ecosystem services can be supported by artificially reconstructed wetlands. Furthermore, the costs of such restoration can be enormous, as for example the US Congressional appropriation of $7.8 billion for the restoration of the Everglades cordgrass system in Florida. As a rule of thumb, most ecologists and policy-makers would agree that it makes more sense to protect it than to lose it and then spend time and money trying to restore it (NRC, 1992). However, restoration zones can and should be a part of the ocean zoning schemes, especially in heavily impacted areas.

Promote applied research

Scientific research should be targeted towards fundamental questions on how marine ecosystems function, what our impact on them really is and what can be done to effectively mitigate against the loss and degradation of these habitats. In order to convince policy-makers and the public that effective protection of marine environments is of paramount importance, investment in better economic valuations of these areas is needed so that the trade-offs that are made when development threatens ocean areas can be better understood and evaluated.

Look for ways to harmonize legislation

Given that most jurisdictional boundaries do not equate to ecological boundaries, there is a real advantage if the legislative arrangements in adjoining marine jurisdictions are complementary (in other words, are close to a 'mirror image' of each other). For example, in the Great Barrier Reef, the fact that a state zoning plan which includes tidal waters has virtually the same zoning provisions as the adjoining federal zoning plan means there is no need to determine exactly where the low water-mark boundary lies. This ensures that it is easier for the public to understand and there is far less onus upon enforcement officers to prove the exact the location of a jurisdictional boundary. A cooperative approach also recognizes the efficiencies to be made through the complementary legislation and the consequent integration of management approaches, such as less duplication and economies of scale arising from the sharing of existing infrastructure and resources.

Placing ocean zoning in a wider context of management

The story of human impacts on marine areas is a complex one involving not only a large number of diverse impacts and drivers behind those impacts, acting simultaneously, but also cumulative effects over time. Unfortunately, we often respond to such impacts on natural systems only after the damage is done – and our response is typically too little, too late. Ocean zoning can certainly help mitigate damage and destruction, but such efforts will only be successful in conservation and resource management if coupled to other management measures.

Thus, regardless of how ocean zoning originates and is implemented, it is imperative that it be embedded in a wider hierarchy of management and that thought be given to the effective governance of that entire hierarchy. Figure 14.1 overleaf illustrates the relationship between zoning and other spatial management initiatives.

An analysis of the efficacy of coastal and marine protected areas, sustainable traditional use regimes and common property management regimes highlights the fact that all such local action must be supplemented by effective management at much larger scales (Agardy, 2008a). Indeed the inter-linkages between terrestrial environments, freshwater, coastal systems and the marine realm prevent local interventions from succeeding unless the larger context is addressed. In order for marine zoning to succeed, management of these areas should address all the direct threats to marine and coastal areas. In most habitats, these threats are multiple and cumulative over time.

A relatively recent movement in this direction is the coupling of coastal zone management with catchment basin or watershed management, as has occurred under the European Water Framework Directive, the Reef Water Quality Protection Plan for the Great Barrier Reef and projects undertaken under the LOICZ (Land-Ocean Interactions in the Coastal Zone) initiative. Such freshwater/marine system coupling has resulted in lower pollutant loads and improved conditions in estuaries. However, due to the fluid nature of the marine system and the large-scale interconnectivities, even larger scale integrated management initiatives are required for effective management of coastal and marine systems over the long term.

Several international instruments provide a framework for such larger scale regional cooperation, including: the 1982 United Nations Convention on the Law of the Sea; UN Regional Seas Conventions and Action Plans; the 1995 Global Programme of Action for the Protection of the Marine Environment from Land-based Activities; the Jakarta Mandate on the Conservation and Sustainable Use of Marine and Coastal Biological Diversity; and the RAMSAR Convention. Yet although global treaties and multilateral agreements can bridge some of the gaps that occur between small-scale interventions on the ground and large-scale coastal problems, most of these international instruments have not been effective in reversing environmental degradation. For shared coastal and marine resources, it may well be that regional agreements allowing for comprehensive ocean zoning will prove more effective, especially when such

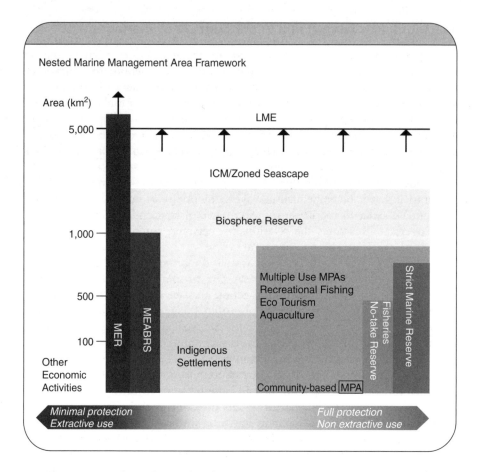

Figure 14.1 *The Relationship between marine zoning and other spatial management initiatives*

Source: Castillo et al, 2006

agreements capitalize on better understandings of costs and benefits accruing from shared responsibilities in conserving the marine environment.

References

Agardy, T. (1997) *Marine Protected Areas and Ocean Conservation*, R. E. Landes Press, Austin, TX, USA

Agardy, T. (2005) 'Global marine policy versus site-level conservation: the mismatch of scales and its implications', Invited paper, *Marine Ecology Progress Series*, vol 300, pp242–248

Agardy, T. (2008a) 'Casting off the chains that bind us to ineffective ocean management', *Ocean Yearbook*, vol 22, pp1–17

Agardy, T. (2008b) 'The marine leap: conservation banking and the brave new world', chapter 11 in N. Carroll, J. Fox and R. Bayon (eds) *Conservation and Biodiversity Banking*, Earthscan, London, pp181–186

Castilla, J. C., P. Christie, J. Cordell and M. Hatziolos (2006) 'Scaling up: Taking marine management beyond MPAs', *World Bank report # 36635-GLB*, Washington, DC, 100pp

Connaughton, J. (2006) pers. comm., Marine Protected Area Federal Advisory Committee meeting deliberations, Washington, DC

Curran, S. and T. Agardy (2002) 'Common property systems, migration, and coastal ecosystems', *Ambio*, vol 31, no 4, pp303–5

Day, J. (2009) 'Governance of the Great Barrier Reef Marine Park', paper prepared for a workshop on MPA Governance, 11–16 October 2009, Lošinj, Croatia

Ehler, C. and F. Douvere (2007) 'Visions for a Sea Change', *Report of the First International Workshop on Marine Spatial Planning. Intergovernmental Oceanographic Commission and Man and the Biosphere Programme*, IOC Manual and Guideas 48, ICAM Dossier no 4, UNESCO, Paris

Gaines, S. and C. Costello (2009) 'Marine protected areas and catch shares – coupling proven tools to ensure a sustainable global seafood supply', ColacMAR meeting, Cuba

Hendrick, D. (2005) 'The new frontier: zoning rules to protect marine resources', *E: The Environmental Magazine*, March–April 2005

Hyrenbach, K. D., K. A. Forney and P. K. Dayton (2001) 'Marine protected areas and the management of pelagic systems', *Aquatic Conservation: Marine and Freshwater Ecosystems*, vol 10, pp437–458

Juda, L. and T. Hennessey (2001) 'Governance profiles and the management and use of large marine ecosystems', *Ocean Development International Law*, vol 32, no 1, pp43–69

Kimball, L. A. (2001) *International Ocean Governance. Using International Law and Organizations to Manage Resources Sustainably*, IUCN, Gland, Switzerland, and Cambridge, UK, 124pp

National Research Council (NRC) (1992) *Restoration of Aquatic Systems: Science, Technology, and Public Policy*, National Academy Press, Washington, DC, 576pp

National Research Council (NRC) (1999) *Sustaining Marine Fisheries*, National Academy Press, Washington, DC

National Research Council (NRC) (2008) *Increasing Capacity for Stewardship of Oceans and Coasts: A Priority for the 21st Century*, National Academy Press, Washington, DC

UNESCO (1996) 'Biosphere Reserves: The Seville Strategy and the Statutory Framework of the World Network', UNESCO, Paris, 1996, 19pp

Villa, F., L. Tunesi and T. Agardy (2001) 'Optimal zoning of a marine protected area: The case of the Asinara National Marine Reserve of Italy', *Conservation Biology*, vol 16, no 2, pp515–526

Young, S. (2006) 'Would ocean zoning be an effective ocean management tool for the waters off the coast of Massachusetts?', *Vermont Journal of Environmental Law*, vol 7, pp1–19

15
Conclusions

T. Agardy

Dhow sailing in the Indian Ocean off Zanzibar, East Africa

Recapping why ocean zoning makes sense

Ocean zoning is no panacea, but it is an extremely powerful tool that can force governments to go beyond endless planning to a marine management system that is strategic, transparent and adaptable.

Ocean zoning improves public understanding and can thereby reduce conflicts. Determining what uses are appropriate where, and what levels of use will keep ecosystems productive and healthy, provides us a much-needed powerful new tool for averting continued degradation and overuse of marine and coastal areas – degradation that has recently been revealed to be far more extensive than anyone had realized (Halpern et al., 2008).

As stated in the introduction to this book (Chapter 1), zoning is simple, straightforward, systematic and strategic. The basic process involves planners identifying ecologically vital areas or areas with a high concentration of ecosystem services, assessing threats to the delivery of those services and developing zoning schemes that prohibit harmful uses while permitting other uses at levels ensuring sustainability. To some extent we already do this when designing marine protected areas, designating closed areas or determining management regimes for certain fisheries. But comprehensive ocean zoning takes this much further – allowing the coordination of piecemeal attempts to manage different resources, different areas and different users into a coherent and strategic whole. If stakeholders are actively involved in the planning process, it is likely that the resulting plans will be welcomed by users and conservationists alike, as they clearly lay out what we know about marine ecosystem value, vulnerabilities and needed management. The greater transparency also creates opportunities for shared stewardship of oceans areas and resources.

Clarification of rights greatly improves our ability to reverse degradation and improve ocean health. Understanding such rights and then coupling them to responsibilities via comprehensive ocean zoning schemes that are overseen by myriad forms of governance will greatly improve our ability to manage impacts and safeguard the seas.

This is because no government is willing or able to shoulder full responsibility for ocean management. Beyond that, there is profit to be made by protecting ecosystem services via ocean zoning. Zoning and payments for ecosystem services markets can and should be coupled – I can envision 'trading zones' where developers buy credits for wetlands protection from environmental groups or private landowners, or where the insurance industry invests in barrier beach protection in order to minimize their own risks (Agardy, 2008).

The possibilities are endless – and already well-developed on land.

Taking the leap: moving from piecemeal to comprehensive ocean zoning

Comprehensive ocean zoning is only a tool, but a potentially powerful tool at that. In fact, I would argue that ocean zoning may be the single most potent advance in marine conservation in modern times, representing a radical departure from the piecemeal and ineffective sectoral management we have practised to date.

Planning and implementing ocean zoning across large geographic scales forces us to do many useful things – things not only necessary for the zoning process but also beneficial for the application of other management tools. Among these are: 1) harnessing the best available information to pinpoint the ecologically most critical or sensitive areas within a region, 2) understanding and displaying uses of the marine and coastal environment, as well as jurisdictions of various management and protection authorities, and 3) communicating information about the relative value of goods and services produced by natural ecosystems in any given area.

Assessing the ecological importance of different areas as well as their other values, and at the same time focusing attention on rights and responsibilities, can only enhance any attempt to manage the oceans and coasts. Much like the power of mathematical models forces us to identify and quantify parameters, so ocean zoning forces us to recognize multiple uses, jurisdictions and rights, and cumulative impacts over time. This is critically important, because not having a clear understanding of the impacts of myriad uses on marine and coastal ecosystems has stymied our attempts to manage in the past. And while coastal zone management (ICM or ICZM) has worked to overcome this hurdle on the land side of the coastal zone, planning for ocean zoning can force such holistic and comprehensive understanding across biomes and at much wider scales.

The comprehensive ocean zoning process catalyses the gathering of such information, but it does more than that: zoning requires we take the information we have, on ecology and environment, on human uses and value, and on impact and sustainable limits of use, and map it. Graphically portraying such information can bring some clarity to what is sometimes an impossibly complex set of management problems. And maps can also help raise public awareness about the diversity of values that marine systems support and, by causing the public to understand how complex a task marine management really is, can generate public support for effective management regimes.

Mapping serves a social function that goes beyond just relaying information. Sociologists have long used community mapping projects to understand perceptions about communal resources and about the sometimes unstructured social institutions that govern the use of such resources. Community mapping and tenure mapping are not restricted to rural or underdeveloped areas; the NGO Finding Sanctuary in the United Kingdom, for instance, has initiated a project called FisherMap that taps the knowledge fishers have to do participatory planning. Initiatives such as FisherMap facilitate stakeholder engagement – drawing users and other stakeholders into a bona fide participatory planning process and thereby sowing the seeds of stewardship (des Clers et al., 2008).

The importance of this stakeholder identification step cannot be overemphasized (Doherty and Butler, 2006). Many a management regime has failed because the planning of management was perceived as top-down, or exclusionary, and the backlash once the regulations were imposed was so great that the management regime collapsed. This is the case time and time again with marine protected areas that are decreed without adequate participation of local communities and user groups; similarly the formulation of place-based fisheries management measures is put at risk when fishers feel they have been excluded from the process. Conversely, some of the best integrated coastal management initiatives are those that have been developed with, and essentially by, local communities, who see themselves not so much as objects of management but as stewards for the ecosystem. However, when one moves out to sea, knowing who to engage as stakeholders, and creating a sense of stewardship among users, are more difficult.

Bonnie McCay, an anthropologist at Rutgers University, is undaunted and sees real opportunity in community-based efforts at sea. In a keynote address to the MARE Conference (the 2007 'People and the Sea' conference, sponsored by the Centre for Maritime Research at the University of Amsterdam), McCay described how understanding the seaward boundaries around what is considered the coastal zone also helps people see the resources of the sea and coasts not as the property of no one (*res nullius*) but rather as a common property (*res communis*). This perception of commons, even in the face of open access, can then lead to growth of 'community'. In her words:

> *Finally, open access indeed sets up the potential for the dynamic that leads to 'tragedies of the commons', but it can also be one that has the potential for 'comedies of the commons' ... or situations in which some of those involved work together to come up with rules and restrictions, restrictions on what one does rather than on whether one has access. The inclusive, open access commons can be a managed commons, as most easily seen in the case of public parks.*
>
> *Bridging, conflating, and sometimes expanding private and public interests to something approaching what we are calling 'common property', that 'in-between' condition of sharing and community, people are engaged in 'comedies of the commons'. In a world of institutions, people have and sometimes take the opportunity to review, reflect, and revise the situation, to create rules, adapt old ones, and try to make them work, and this is what I mean by 'comedy', following ancient Greek drama, where comedy is '[T]he drama of humans as social rather than private beings, a drama of social actions having a frankly corrective purpose'.* (McCay, 2007, p9)

Thus it is possible that the information gathering and the displaying of that information as is required for the zoning process can be like the glue that holds stakeholders together and elicits among them a forming of institutions to oversee their use of ocean space and resources, even in the absence of a formal (top-down or bureaucratized) governance regime. This is a spin-off benefit of zoning that can then be built upon, as management regimes for each zone are codified. In other words, a zoning plan could have not only no-take or no-entry zones to protect the ecologically most critical or vulnerable places or habitats, but also community zones, in which use is managed, or co-managed, by a community institution – one that may have emerged as a community institution through the process of planning the zonation.

One can take this farther, fully into the realm of public-private partnerships, discussing the link between ocean zoning and the development of payments for ecosystem services (PES) markets. Such markets reward the stewards of ecosystems for protecting the delivery of services that such ecosystems provide. Beneficiaries of those services – industry, individuals or government as representing the public – pay for the protection of the services from which they themselves benefit.

The infusion of private sector financing into what conventionally are publically managed areas provides much needed additional income for costly management activities, and it allows government to share the burden of management with local communities and users, who may indeed be better prepared to monitor and manage coastal and marine areas. Zoning can facilitate the creation of these innovative financing systems by elucidating the location of the habitats where these valuable services are being delivered, by clarifying rights and responsibilities (in a geographical sense) and by reducing risk for private investors through that very clarification of use, access and management rights (Agardy, 2008).

Another benefit that zoning brings to the ocean management table is its ability to highlight understudied and underrepresented areas, such as the pelagic environments far offshore, the deep seas, etc. One can envision that in very poorly studied areas (remember, the sea is less well charted and studied than the surface of the moon!) research zones could be established to facilitate targeted research and a better overall understanding of marine ecology and ocean condition.

Finally, comprehensive ocean zoning allows us to be integrative and strategic like no other management tool can. It requires understanding of linkages – between different coastal and marine habitats, and between terrestrial/freshwater and marine habitats, as well as between uses of resources and space and subsequent impacts on ecology. By taking such linkages into account in planning management, zoning can maximize the efficiency of management. And when zonation is planned by evaluating different scenarios and choosing desirable outcomes, zoning can be inherently strategic.

Who will lead the process?

If zoning is imposed from the top without consultation, there will be a huge backlash. We know this from examples such as the national zoning of Belgium's EEZ, as well as the progress being made in the states of California and Massachusetts to zone state waters. But if zoning plans are developed through a participatory process (avoiding high-tech decision-support tools which when not explained to the public cause great suspicion), zoning will be understood and accepted. We already do zoning on land with little angst and have tested its application at sea through marine protected areas such as the Great Barrier Reef Marine Park. But comprehensive ocean zoning will require planning on scales hitherto shunned, incorporating not only large marine ecosystems or regional seas but also adjacent (and ecologically linked) watersheds.

Thus, true leadership is needed to institute comprehensive ocean zoning. Ocean agencies within national governments should lead the process, setting up regional groups comprised of representatives of all affected agencies, industries and communities to develop zoning plans at workable scales. All institutions involved in governance and marine management would do well to heed the lessons learned from those forging ahead with small-scale or specialized ocean zoning plans, many of which have been captured in the pages of this book. Whether they harness comprehensive ocean zoning for ecosystem-based fisheries management, link ocean zoning with the development of ecosystem services markets, and/or explore opportunities for public-private partnerships – to set out on such unpredictable (though invariably fruitful) journeys takes real vision.

Final thoughts

The forward movement towards more strategic approaches to managing marine use and conserving marine ecosystems and their services is happening all around the world. Much of this forward movement has been facilitated by an integrated, systematic and hierarchical approach that allows nations to address various geographic scopes and scales of marine management issues simultaneously and in a more holistic manner. By using large marine regions (ecoregions, regional seas, semi-enclosed seas, etc.) as the focus of management for developing comprehensive ocean zoning, the actors involved in governance of marine areas and uses agencies can most effectively come together to address the full spectrum of threats.

Marine management goals such as biodiversity conservation, conservation of rare and threatened species, maintenance of natural ecosystem functioning at a regional scale, management of fisheries, recreation, reducing conflict, and promoting education and research could be better addressed through comprehensive ocean zoning undertaken at large or regional scales. The integrated approach inherent in zoning is a natural response to a complex set of ecological processes and environmental problems and is an efficient way to allocate scarce time and resources to combating the issues that parties deem to be most critical. Nations and/or agencies that participate reap the benefits of more effective conservation, while bearing fewer costs by spreading management costs widely and by taking advantage of economies of scale in management training, enforcement, scientific monitoring and the like (Agardy, 2005). What is needed is leadership that understands and acknowledges the big picture, and has the strength of conviction to move a complicated agenda forward.

Ocean zoning will undoubtedly gain in popularity as further tests of the concept emerge at various scales and as pressures on and conflicts in the sea increase. Pioneers in implementing ocean zoning will take many different courses and tacks, and some may end up floundering, but the great potential of ocean zoning suggests the journeys – like all good adventures – will turn out to have been well worth the trouble.

References

Agardy, T. (2005) 'Global marine conservation policy versus site level implementation: the mismatch of scale and its implications', pp242–248 in H. I. Browman, K. I. Stergiou (eds) 'Politics and socio-economics of ecosystem-based management of marine resources', Journal Special Issue, *Marine Ecology Progress Series*, vol 300, pp241–296.

Agardy, T. (2008) 'The marine leap: conservation banking and the brave new world', chapter 11 in N. Carroll, J. Fox and R. Bayon (eds) *Conservation and Biodiversity Banking*, Earthscan, London, pp181–186

des Clers, S., S. Lewin, D. Edwards, S. Searle, L. Lieberknecht and D. Murphy (2008) 'FisherMap – mapping the Grounds: recording fishermen's use of the seas', Final report of Finding Sanctuary, UK

Doherty, P. A. and M. Butler (2006) 'Ocean zoning in the Northwest Atlantic', *Marine Policy*, vol 30 (2006), pp389–391

Halpern B. S., S. Walbridge, K. A. Selkoe, C. V. Kappel, F. Micheli, C. D'Agrosa, J. F. Bruno, K. S. Casey, C. Ebert, H. E. Fox, R. Fujita, D. Heinemann, H. S. Lenihan, E. M. P. Madin, M. T. Perry, E. R. Selig, M. Spalding, R. Steneck and R. Watson (2008) 'A global map of human impact on marine ecosystems', *Science*, vol 319, pp948–952

McCay, B. (2007) 'The littoral and the liminal: challenges to the management of the coastal and marine commons', *MAST*, 7.1

Annex:
IUCN Protected Area Categories

The following background is from the National Research Council (2001) publication: *Marine Protected Areas*, National Academy Press, Washington, DC. Protected Area categories are from 'Guidelines for Protected Area Management Categories' (IUCN, 1994).

The definition of a marine protected area (MPA) adopted by IUCN and other international and national bodies is:

> *Any area of intertidal or subtidal terrain, together with its overlying water and associated flora, fauna, historical and cultural features, which has been reserved by law or other effective means to protect part or all of the enclosed environment.* (Kelleher and Kenchington, 1992)

There are several important features of the IUCN categorization scheme that it is important to note. They are:

- the basis of categorization is by primary management objective
- assignment to a category is not a commentary on management effectiveness
- the categories system is international
- national names for protected areas of the same category vary
- all categories are important
- though the primary objective of an MPA will determine the category, the MPA may contain zones which have other objectives. However, for the purpose of categorization, at least three-quarters of the MPA must be managed for the primary objective and the management of the remaining area must not conflict with that primary objective

Category I

Strict Nature Reserve – Wilderness Area: Protected Area Managed Mainly for Science or Wilderness Protection

Category IA

Strict Nature Reserve: Protected Area Managed Mainly for Science

Definition
Area of land and/or sea possessing some outstanding or representitive ecosystems, geological or physiological features and/or species, available primarily for scientific research and/or environmental monitoring.

Objectives of management
- to preserve habitats, ecosystems and species in as undisturbed a state as possible
- to maintain genetic resources in a dynamic and evolutionary state
- to maintain established ecological processes
- to safeguard structural landscape features or rock exposures
- to secure examples of the natural environment for scientific studies, environmental monitoring and education, including baseline areas from which all avoidable access is excluded
- to minimize disturbance by careful planning and execution of research and other approved activities
- to limit public access

Guidance for selection
- The area should be large enough to ensure the integrity of its ecosystems and to accomplish the management objectives for which it is protected
- The area should be significantly free of direct human intervention and capable of remaining so
- The conservation of the area's biodiversity should be achievable through protection and not require substantial active management or habitat manipulation (see Category IV)

Organizational responsibility
Ownership and control should be by the national or other level of government, acting through a professionally qualified agency, or by a private foundation, university or institution which has an established research or conservation function, or by owners working in cooperation with any of the foregoing government or private institutions. Adequate safeguards and controls relating to long-term protection should be secured before designation. International agreements over areas subject to disputed national sovereignty can provide exceptions (e.g., Antarctica).

Equivalent category in 1978 System
Scientific Research/Strict Nature Reserve.

Category IB

Wilderness Area: Protected Area Managed Mainly for Wilderness Protection

Definition
Large area of unmodified or slightly modified land, and/or sea, retaining its natural character and influence, without permanent or significant habitation, which is protected and managed so as to preserve its natural condition.

Objectives of management
- to ensure that future generations have the opportunity to experience understanding and enjoyment of areas that have been largely undisturbed by human action over a long period of time
- to maintain the essential natural attributes and qualities of the environment over the long term
- to provide for public access at levels and of a type which will serve best the physical and spiritual well-being of visitors and maintain the wilderness qualities of the area for present and future generations
- to enable indigenous human communities living at low density and in balance with the available resources to maintain their lifestyle

Guidance for selection
- The area should possess high natural quality, be governed primarily by the forces of nature, with human disturbance substantially absent, and be likely to continue to display those attributes if managed as proposed
- The area should contain significant ecological, geological, physiogeographic, or other features of scientific, educational, scenic or historic value
- The area should offer outstanding opportunities for solitude, enjoyed once the area has been reached, by simple, quiet, non-polluting and non-intrusive means of travel (i.e., non-motorized)
- The area should be of sufficient size to make practical such preservation and use

Organizational responsibility
As for Sub-Category 1A.

Equivalent category
This sub-category did not appear in the 1978 system, but has been introduced following the IUCN General Assembly Resolution (16/34) on Protection of Wilderness Resources and Values, adopted at the 1984 General Assembly in Madrid, Spain.

Category II

National Park: Protected Area Managed Mainly for Ecosystem Protection and Recreation

Definition
Natural area of land and/or sea, designated to (a) protect the ecological integrity of one or more ecosystems for present and future generations, (b) exclude exploitation or occupation inimical to the purposes of designation of the area, and (c) provide a foundation for spiritual, scientific, educational, recreational and visitor opportunities, all of which must be environmentally and culturally compatible.

Objectives of management
- to protect natural and scenic areas of national and international significance for spiritual, scientific, educational, recreational or tourist purposes
- to perpetuate, in as natural a state as possible, representative examples of physiographic regions, biotic communities, genetic resources and species to provide ecological stability and diversity
- to manage visitor use for inspirational, educational, cultural and recreational purposes at a level which will maintain the area in a natural or near-natural state
- to eliminate and thereafter prevent exploitation or occupation inimical to the purposes of designation
- to maintain respect for the ecological, geomorphologic, sacred or aesthetic attributes which warranted designation
- to take into account the needs of indigenous people, including subsistence resource use, in so far as these will not adversely affect the other objectives of management

Guidance for selection
- The area should contain a representative sample of major natural regions, features or scenery, where plant and animal species, habitats and geomorphological sites are of special spiritual, scientific, educational, recreational and tourist significance
- The area should be large enough to contain one or more entire ecosystems not materially altered by current human occupation or exploitation

Organizational responsibility
Ownership and management should normally be by the highest competent authority of the nation having jurisdiction over it. However, they may also be vested in another level of government, council of indigenous people, foundation or other legally established body which has dedicated the area to long-term conservation.

Equivalent category in 1978 System
National Park.

Category III

Natural Monument: Protected Area Managed Mainly for Conservation of Specific Natural Features

Definition
Area containing one, or more, specific natural or natural/cultural features which is of outstanding or unique value because of its inherent rarity, representative or aesthetic qualities or cultural significance.

Objectives of management
- to protect or preserve in perpetuity specific outstanding natural features because of their natural significance, unique or representational quality, and/or spiritual connotations
- to an extent consistent with the foregoing objective, to provide opportunities for research, education, interpretation and public appreciation
- to eliminate and thereafter prevent exploitation or occupation inimical to the purpose of designation
- to deliver to any resident population such benefits as are consistent with the other objectives of management

Guidance for selection
- The area should contain one or more features of outstanding significance (appropriate natural features include spectacular waterfalls, caves, craters, fossil beds, sand dunes and marine features, along with unique or representative fauna and flora; associated cultural features might include cave dwellings, cliff-top forts, archaeological sites or natural sites which have heritage significance to indigenous peoples).
- The area should be large enough to protect the integrity of the feature and its immediately related surroundings.

Organizational responsibility
Ownership and management should be by the national government or, with appropriate safeguards and controls, by another level of government, council of indigenous people, non-profit trust, corporation or, exceptionally, by a private body, provided the long-term protection of the inherent character of the area is assured before designation.

Equivalent category in 1978 System
Natural Monument/Natural Landmark.

Category IV

Habitat/Species Management Area: Protected Area Managed Mainly for Conservation through Management Intervention

Definition   Area of land and/or sea subject to active intervention for management purposes so as to ensure the maintenance of habitats and/or to meet the requirements of specific species.

Objectives of management
- to secure and maintain the habitat conditions necessary to protect significant species, groups of species, biotic communities or physical features of the environment where these require specific human manipulation for optimum management
- to facilitate scientific research and environmental monitoring as primary activities associated with sustainable resource management
- to develop limited areas for public education and appreciation of the characteristics of the habitats concerned and of the work of wildlife management
- to eliminate and thereafter prevent exploitation or occupation inimical to the purposes of designation
- to deliver such benefits to people living within the designated area as are consistent with the other objectives of management

Guidance for selection
- The area should play an important role in the protection of nature and the survival of species (incorporating, as appropriate, breeding areas, wetlands, coral reefs, estuaries, grasslands, forests or spawning areas, including marine feeding beds)
- The area should be one where the protection of the habitat is essential to the well-being of nationally or locally important flora, or to resident or migratory fauna
- Conservation of these habitats and species should depend upon active intervention by the management authority, if necessary through habitat manipulation (see Category IA)
- The size of the area should depend on the habitat requirements of the species to be protected and may range from relatively small to very extensive

Organizational responsibility
Ownership and management should be by the national government or, with appropriate safeguards and controls, by another level of government, non-profit trust, corporation, private group or individual.

Equivalent category in 1978 System   Nature Conservation Reserve/Managed Nature Reserve/Wildlife Sanctuary.

Category V

Protected Landscape/Seascape: Protected Area Managed Mainly for Landscape/Seascape Conservation and Recreation

Definition
Area of land, with coast and sea as appropriate, where the interaction of people and nature over time has produced an area of distinct character with significant aesthetic, ecological and/or cultural value, and often with high biological diversity. Safeguarding the integrity of this traditional interaction is vital to the protection, maintenance and evolution of such an area.

Objectives of management
- to maintain the harmonious interaction of nature and culture through the protection of landscape, and/or seascape and the continuation of traditional land uses, building practices and social and cultural manifestations
- to support lifestyles and economic activities which are in harmony with nature and the preservation of the social and cultural fabric of the communities concerned
- to maintain the diversity of landscape and habitat, and of associated species and ecosystems
- to eliminate where necessary, and thereafter prevent, land uses and activities which are inappropriate in scale and/or character
- to provide opportunities for public enjoyment through recreation and tourism appropriate in type and scale to the essential qualities of the areas
- to encourage scientific and educational activities that will contribute to the long term well-being of resident populations and to the development of public support for the environmental protection of such areas
- to bring benefits to, and to contribute to the welfare of, the local community through the provision of natural products (such as forest and fisheries products) and services (such as clean water or income derived from sustainable forms of tourism)

Guidance for selection
- The area should possess a landscape and/or coastal and island seascape of high scenic quality, with diverse associated habitats, flora and fauna along with manifestations of unique or traditional land-use patterns and social organizations as evidenced in human settlements and local customs, livelihoods and beliefs
- The area should provide opportunities for public enjoyment through recreation and tourism within its normal lifestyle and economic activities

Organizational responsibility
The area may be owned by a public authority, but is more likely to comprise a mosaic of private and public ownerships operating a variety of management

regimes. These regimes should be subject to a degree of planning or other control and supported, where appropriate, by public funding and other incentives, to ensure that the quality of the landscape/seascape and the relevant local customs and beliefs are maintained in the long term.

Equivalent category in 1978 System
Protected Landscape.

Category VI

Managed Resource Protected Area: Protected Area Managed Mainly for the Sustainable Use of Natural Ecosystems

Definition   Area containing predominantly unmodified natural systems, managed to ensure long term protection and maintenance of biological diversity, while providing at the same time a sustainable flow of natural products and services to meet community needs.

Objectives of management
- to protect and maintain the biological diversity and other natural values of the area in the long term
- to promote sound management practices for sustainable production purposes
- to protect the natural resource base from being alienated for other land-use purposes that would be detrimental to the area's biological diversity
- to contribute to regional and national development

Guidance for selection
- The area should be at least two-thirds in a natural condition, although it may also contain limited areas of modified ecosystems; large commercial plantations would not be appropriate for inclusion
- The area should be large enough to absorb sustainable resource uses without detriment to its overall long-term natural values

Organizational responsibility
Management should be undertaken by public bodies with an unambiguous remit for conservation, and carried out in partnership with the local community; or management may be provided through local custom supported and advised by governmental or non-governmental agencies. Ownership may be by the national or other level of government, the community, private individuals or a combination of these.

Equivalent category in 1978 System
This category does not correspond directly with any of those in the 1978 system, although it is likely to include some areas previously classified as 'Resource

Reserves', 'Natural Biotic Areas/Anthropological Reserves' and 'Multiple Use Management Areas/Managed Resource Areas'.

Source: Kelleher, G. and R. Kenchington (1992) *Guidelines for Establishing Marine Protected Areas*, Marine Conservation and Development Report, IUCN, Gland, Switzerland; IUCN (1994) *Guidelines for Protected Areas Management Categories*, IUCN, Cambridge, UK and Gland, Switzerland

Recommended Further Reading

Maria Agardy

The author in the Galapagos Islands Marine Park

Agardy, T. (1997) *Marine Protected Areas and Ocean Conservation*, R. E. Landes Press, Austin, TX, USA

Agardy, T. (2005) 'Global marine conservation policy versus site level implementation: the mismatch of scale and its implications', pp242–248 in H. I. Browman and K. I. Stergiou (eds) 'Politics and socio-economics of ecosystem-based management of marine resources', special journal issue, *Marine Ecology Progress Series*, vol 300, pp241–296

Agardy, T. (2008) 'Casting off the chains that bind us to ineffective ocean management', *Ocean Yearbook*, vol 22, pp1–17

Agardy, T. (2008) 'The marine leap: conservation banking and the brave new world', chapter 11 in N. Carroll, J. Fox, and R. Bayon (eds) *Conservation and Biodiversity Banking*, Earthscan, London, pp181–186

Crowder, L., G. Osherenko, O. R. Young, S. Airame, E. A. Norse and N. Baron (2006) 'Sustainability – resolving mismatches in US ocean governance', *Science*, vol 313, no 5787, pp617-618

Day, J. C. (2002) 'Zoning – Lessons from the Great Barrier Reef Marine Park', *Ocean and Coastal Management*, vol 45, pp139–156

Day, J. C. (2006) 'Marine protected areas', pp603–634 in M. Lockwood, H. Worboys and A. Kothari (eds) *Managing Protected Areas: A Global Guide*, Earthscan, London

Day, J. C. (2008) 'EBM perspective: clarifying misconceptions about zoning – the GBR Example', *Marine Ecosystems and Management (MEAM)*, vol 2, no 1, p6

Derous, S., T. Agardy, H. Hillewaert, K. Hostens, G. Jamieson, L. Lieberknecht, J. Mees, I. Moulaert, S. Olenin, D. Paelinck, M. Rabaut, E. Rachor, J. Roff, E. W. M. Stienen, J. T. Van der Wal, V. Van Lancker, E. Verfaillie, M. Vinc, J. M. Weslawski and S. Degraer (2007) 'A concept for biological valuation in the marine environment', *Oceanologia*, vol 49, no 1, pp99–128

Doherty, P. A. and M. Butler (2006) 'Ocean zoning in the Northwest Atlantic', *Marine Policy*, vol 30, pp389–391

Douvere, F., F. Maes, A. Vanhulle and J. Schrijvers (2006) 'The role of marine spatial planning in sea use management: the Belgian case', *Marine Policy*, vol 31, no 2, pp182–191

Ehler, C. and F. Douvere (2007) 'Visions for a Sea Change', *Report of the First International Workshop on Marine Spatial Planning, Intergovernmental Oceanographic Commission and Man and the Biosphere Programme*, IOC Manual and Guides 48, ICAM Dossier no 4, UNESCO, Paris

Ehler, C. and F. Douvere (2009) 'Marine spatial planning: A step-by-step approach toward ecosystem-based management', *Intergovernmental Oceanographic Commission and Man and the Biosphere Programme*, IOC Manual and Guides 53, ICAM Dossier no 6, UNESCO, Paris

Eilperin, J. (2009) 'Finding space for all in our crowded seas', *Washington Post*, 4 May

European Commission (EC) (2008) 'Roadmap for Maritime Spatial Planning: Achieving Common Principles in the EU Brussels', 25.11.2008 COM(2008) 791 Final

Fernandes L, J. Day, A. Lewis, S. Slegers, B. Kerrigan and D. Breen (2005) 'Establishing representative no-take areas in the Great Barrier Reef: large-scale implementation of theory on marine protected areas', *Conservation Biology*, vol 19, no 6, pp1733–1744

Halpern, B. S., K. L. McLeod, A. A. Rosenberg and L. B. Crowder (2008) 'Managing for cumulative impacts in ecosystem-based management through ocean zoning', *Ocean and Coastal Management*, vol 51, pp203–211

Hendrick, D. (2005) 'The new frontier: zoning rules to protect marine resources', *E: The Environmental Magazine*, March–April

Kappel, C., B. S. Halpern, R. G. Martone, F. Micheli, K. A. Selkoe (2009) 'In the zone: Comprehensive ocean protection', *Issues in Science and Technology*, Spring, pp33–44

Kay, R. and J. Alder (2005, 2nd edition) *Coastal Planning and Management*, Taylor and Francis, Abingdon, UK, and New York

Kimball, L. A. (2001) *International Ocean Governance. Using International Law and Organizations to Manage Resources Sustainably*, IUCN, Gland, Switzerland, and Cambridge, UK, 124pp

MEAM (2008) 'Comprehensive ocean zoning: Answering questions about this important tool for EBM', *Marine Ecosystems and Management (MEAM)*, vol 2, no 1, pp1–4

National Research Council (NRC) (2008) *Increasing Capacity for Stewardship of Oceans and Coasts: A Priority for the 21st Century*, National Academy Press, Washington, DC

Norse, E. A. (2005) 'Ending the range wars on the last frontier: zoning the sea', in E. A. Norse and L. B. Crowder (eds) *Marine Conservation Biology: The Science of Maintaining the Sea's Biodiversity*, Island Press, Washington, DC, pp422–443

Sanchirico, J. (2009) 'Better defined rights and responsibilities in marine adaptation policy', *Resources for the Future Issue Brief*, RFF, Washington, DC

Villa, F., L. Tunesi and T. Agardy (2001) 'Optimal zoning of a marine protected area: the case of the Asinara National Marine Reserve of Italy', *Conservation Biology*, vol 16, no 2, pp515–526

Young, S. (2006) 'Would ocean zoning be an effective ocean management tool for the waters off the coast of Massachusetts?', *Vermont Journal of Environmental Law*, vol 7, pp1–19

Index

access to coastal regions and resources 4, 194
  indigenous peoples 31–32, 81–82
  United Kingdom 89
adaptive management 35, 43, 48, 137, 172
  baselines 107
  examples of 145, 166
  research sites to support 52–53
Africa 145, 168
  East 139–142
  South 142–145
  West 134–138
algal blooms 4
applied research 186
aquaculture, New Zealand 76, 78, 81–82
Arctic Ocean, closure to trawl fishing 34
Areas Beyond National Jurisdictions (ABNJs) 122, 124–129
Asinara Marine National Park 112–113, 118–119, 168
  choosing preferred option 116–118
  developing zoning options 113–116
Australia 8, 32, 71
  see also Great Barrier Reef Marine Park (GBRMP)
Azores meeting 125–127

Banco Chinchorro Reserve 180
Barale, Vittorio 44
Barcelona Convention 11, 122
  Regional Activity Centre for Specially Protected Areas (RAC/SPA) 125, 127, 129
Barents Sea 106–107
Belgium 8, 98, 101–102, 167–168

biological valuation 18, 36, 51, 102–104
  GAUFRE project 104–105
  Marine Spatial Planning Policy Framework 105–106
benefits of ocean zoning 7, 15, 192, 194–195
Benguela Current Commission (BCC) 143
Benguela Current Large Marine Ecosystem Project 143, 145
Benn, Hilary 89, 94
benthic/pelagic coupling 12, 48–49
benthic protection areas (BPAs) 79
Bijagos Biosphere Reserve 135–136
biocenotic (habitat) mapping 113
biological valuation 18, 36, 51, 102–104
biophysical operating principles, Great Barrier Reef Marine Park 64, 65
bioregionalism 14
Biosphere Reserve programme 18, 34, 71, 177–178
Blueprint 2050 142
British Columbia 148, 168–169
  biogeography 148–150
  spatial planning 150–154
British Columbia Marine Conservation Analysis (BCMCA) 151–153

California 178
  Channel Islands National Marine Sanctuary (CINMS) 28–29
  California Marine Life Protection Act 8, 158–159
Canada 178
  Oceans Act 8, 36, 150
  see also British Columbia

Canary Current Large Marine Ecosystem
    Project (CCLME) 138
cetaceans 124, 131
Channel Islands National Marine
    Sanctuary (CINMS) 28–29
China 8, 182
Chumbe Island Coral Park (CHICOP)
    141–142
civil society governance 179–180
  climate change impacts 3–4, 12, 24
  coastal management initiatives 25
coastal zones
  climate change impacts 3–4
  defining 92–93
  habitat alteration 2–3
  increasing populations and pressure
      on 4–5
  ineffectiveness of management 5–6
Colombia 70, 71–72
co-management 180, 183
commons 34, 35, 44, 181, 183, 194
community mapping 193
competition in fisheries 29, 30, 31
Comprehensive Ocean Zoning (COZ)
    10, 13, 14, 167
  Belgium 102–106
  leadership 195–196
  potential for halting ocean decline
      16–17, 192–193
  scale of management responses 26
  see also implementing the plan; process
      of ocean zoning
concordance/discordance analysis
    114–118
conflict
  fisheries displacement 29, 30
  New Zealand 81–82
  resolution mechanisms 32, 160
connections 11
Convention for the Protection of the
    Marine Environment and the
    Coastal Region of the Mediterranean
    see Barcelona Convention
Coralina 71, 72
coral reefs see Great Barrier Reef Marine
    Park (GBRMP)
corridors 7
criteria for assessment of marine areas
    47–48, 103–104
  Mediterranean Sea 125–126, 127, 128

culture 31

Day, Jon 63, 68
dead zones 4
deep water mining 79
DEFRA (UK Department of
    Environment, Food and Rural
    Affairs) 86–87, 88–89, 92–93
Delphic methods 47, 48, 127, 170
Derous, S. et al. 102–104
displacement of fisheries 26–27
  ecological consequences 30
  economic consequences 27–29
  into new types of fisheries 30–31
  lack of evidence of costs 32
  perceptions and support for marine
      management 31–32
  social consequences 29–30
Dobbin, James 14
dredging 186
Dutch National Water Plan 107–108

easements 10
East Africa 139–142
ecological criteria 48, 51, 125–126, 128
Ecologically or Biologically Significant
    Areas (EBSAs) 125–126, 127, 168
ecological networks 7, 54
ecosystem-based management (EBM) 15
  British Columbia 151
  fisheries 185
  Massachusetts 156–157
  Namibia 142, 143, 145
  Norway 106
  RAMPAO initiative 137–138
  Tanzania 142
ecosystem services xii, 2
  economic value 181, 184, 192, 195
education and awareness 54, 183–184
Ehler, C. and Douvere, F. 13–14, 15–16,
    42–43, 98–99, 105
enforcement 32, 119, 183
  monitoring compliance 179
environmental quality objectives
    (EcoQOs) 107
estuarine habitats 11, 12, 25
European Union (EU)
  Common Fisheries Policy 106
  integration 93
  legislation and initiatives 99

LIFE-Nature project 124
Natura 2000 105
principles for marine spatial planning 12–13, 169, 170–171
Water Framework Directive 11, 25, 94, 187
Exclusive Economic Zones (EEZs)
    Belgium 101–106
    China 182
    Germany 108
    Namibia 145
    New Zealand 36, 76–77, 82
    Tanzania 140
existing use patterns 51

First Nations 148, 153–154, 168
Fisheries Act 1996 (NZ) 78
fisheries exclusion zones 28
fisheries management 2, 5, 185
    Arctic ocean 34
    Mediterranean Sea 34, 124
    Namibia 143
    New Zealand 78
    Norway 98, 106–107
    potential for ocean zoning 14, 182
    Tanzania 140–141
    United Kingdom 90
    see also displacement of fisheries
fishing the line 29
fish stock depletion 2
flagship species 55
Foreshore and Seabed Act (NZ) 81
freshwater/marine system coupling
    see watershed and coastal zone management linkage
funding and revenues 5, 138, 140–141, 178
    community development banks 179
    private sector 141–142, 181, 195

GAUFRE project 105
General Fisheries Commission for the Mediterranean (GFCM) 34, 124
Geographic Information Systems (GIS) 46, 49, 87
    Asinara Marine National Park 113, 114, 116
    when and how to use 169, 171–172
Georges Bank 29, 31
Germany 8, 98, 108

global oceans
    climate change impacts 4, 12
    decline in health 2, 16–17
Global Programme of Action for the Protection of the Marine Environment from Land-based Activities 24–25
goals and objectives for zoning 15, 72, 119, 196
goal-setting 43–44
Great Barrier Reef Marine Park 60, 66–67
governance 12, 18–19, 176–177
    by government 50, 177–178
    civil society in partnership with government 179–180
    special legislation 15, 182–183
Great Barrier Reef Marine Park (GBRMP) 48, 60–61, 69, 918
    existing use patterns 51
    green sea turtle 53
    legal framework for zoning 66–68, 182–183
    lessons derived from 166–167, 171–172
    misconceptions 68–69
    Representative Areas Program 62–64
    rezoning process 64–66
    zone designations and permitted activities 61–62, 63
Great Barrier Reef Marine Park Authority (GBRMPA) 61, 65, 66, 167, 182–183
green zones 63
Guinea Bissau 135–136

habitat alteration 2–3
habitat protection see marine protected areas (MPAs)
Haida Nation 154
Hawaii longline fishery 33, 53
Helsinki Commission (HELCOM) 102, 108
hierarchical planning 11–12, 19, 25, 128
    Irish Sea Pilot 87
high seas MPAs 33, 53–54
    Pelagos Sanctuary 54, 122, 130–131
    see also Areas Beyond National Jurisdictions (ABNJs)
human health 4

human networks 7, 54, 136

ICRAM (Central Institute for Scientific and Technical Marine Advice) 112, 113
implementing the plan 50–51, 176
  governance 18–19, 176–182
  management 176, 180, 183–186, 193–196
  wider context 187–188
  see also legislative frameworks
indigenous peoples 31
  British Columbia 148, 153–154
  New Zealand 81–82
Individual Transferable Quotas (ITQs) 78
information gathering and analysis 15
integrated approaches 12–13, 14, 19, 196
  British Columbia 152–153
  Great Barrier Reef Marine Park 69
  implementing the plan 50, 176
  land and sea-use management 184
  Massachusetts 156–157
  Namibia 145
  New Zealand 78
  and scale 25–26
  Seaflower MPA 71
  Tanzania 142
  United Kingdom 89–91, 92–94
Intergovernmental Oceanographic Commission 9, 98–99
Interim Framework for Effective Coastal and Marine Spatial Planning (US) 160–161
international agreements 98–99
international instruments 24–25, 187–188
International Union for the Conservation of Nature (IUCN), Protected Area categories 199–207
Irish Sea Conservation Zones (ISCZ) project 88–89
Irish Sea Pilot 84–88, 167
Italy see Asinara Marine National Park

Jakarta Mandate on the Conservation and Sustainable Use of Marine and Coastal Biological Diversity 25
Johnson, David 55

Joint Nature Conservation Committee (JNCC) 86–87, 89

Karibouye, Charlotte 138
Kay, R. and Alder, J. 61
keystone species 55

Land-Ocean Interactions in the Coastal Zone (LOICZ) initiative 11, 25, 187
land-use planning 10–11
  see also watershed and coastal zone management linkage
leadership 44, 195–196
legislative frameworks 15, 182–183, 186
  New Zealand 76, 77–78
  property and use rights 26
  zoning in Great Barrier Reef Marine Park 66–68, 186
listed species 33, 53, 112
Living Oceans Society 151
Lofoten Islands 106–107
loggerhead turtles 33, 53, 112
LOICZ (Land-Ocean Interactions in the Coastal Zone) initiative 11, 25, 187

McCay, Bonnie 194
Mafia Island Marine Park (MIMP) 139–141, 182
mapping 42, 49, 193
  BCMCA 152–153
  biocenotic (habitat) 113
  biological valuation 102–104
  concordance 115–117
  GAUFRE project 105
  Irish Sea Pilot 88
Marine and Coastal Access Act 2009 (UK) 84, 86, 89–91, 94, 167
marine conservation 7–8, 33, 53, 55, 112
  scale of response needed 18, 19, 24, 47, 54
  see also marine protected areas (MPAs)
Marine Conservation Society (UK) 91–92
marine ecosystems xii
  degradation 2, 12
  dynamic nature 32, 33, 35, 53
  identification of priority areas 49, 124–128
  scale 18, 24, 53

Marine Legacy Fund (Tanzania) 140–141
Marine Management Organisation (MMO) 84, 88, 89, 90–91
marine policy 24–25
marine protected area (MPA) networks 6–7, 47, 54–56, 88–89
  California 158–159
  planning for the Mediterranean 125–128
  West Africa 136–138
marine protected areas (MPAs) 18, 24
  benefits 52–54
  displacement of effort 30
  high seas *see* high seas MPAs
  identification of priority areas 46–47, 91–92, 124–128
  implications of zoning within 69–72
  IUCN categorization scheme 199–207
  Lundy Island 90, 92
  management effectiveness 119, 185
  Mediterranean Sea 54, 122, 124–128, 129–131
  Namibian Islands 143
  New Zealand 76
  perceptions of displacement 31–32
  spillover 27–28, 29, 30, 182
  starting point for zoning 51–52, 89, 129
  use of zoning 33, 34, 117–119
  see also Asinara Marine National Park; Great Barrier Reef Marine Park (GBRMP)
marine spatial planning/management *see* maritime spatial planning (MSP)
maritime spatial planning (MSP) 6–9, 42
  best practices 46
  British Columbia 150–154
  differences with ocean zoning 13–16
  European Commission principles 12–13, 169, 170–171
  Great Barrier Reef Marine Park 67–68
  Mediterranean initiatives 122–127
  need for user group acceptance 31–32, 44
  North East Atlantic 84–91, 102, 104–108
  policy drivers in Europe 98–101
  US Federal policy 161–163
markets 181, 195
Martin, T.G. et al. 50

MARXAN 46, 48, 49–50, 65
  British Columbia 151, 152
  Irish Sea Pilot 87, 88
  when and how to use 169, 171–172
Massachusetts 8
  Ocean Act 14–15, 156–158
Mediterranean Sea 122–127, 168
  Pelagos Sanctuary 33, 54, 122, 130–131
  restriction of bottom trawling 34, 124
  see also Asinara Marine National Park
Mekong River Commission 178
Merritt Island National Wildlife Refuge 28
methodologies *see* planning tools
Mexico 8
migration corridors 55, 137
MMO (Marine Management Organisation) 84, 88, 89, 90–91
monitoring and evaluation 184
multicriteria evaluation 114
Multipurpose Marine Cadastre (MMC) 161

Namibia 142–145
Namibian Islands MPA 143
Namib Naukluft National Park 143
National Center for Ecological Analysis and Synthesis (NCEAS) 9
National Marine Fisheries Service (US) 33, 53
Natura 2000 Network 105
Nature Conservancy (TNC) 46
Netherlands 98, 107–108
New Zealand 8, 18, 32, 76, 167
  Exclusive Economic Zone (EEZ) 36, 76–77, 82
  fisheries and ocean zoning 78–80
  international marine protection 80–81
  MPA networks 55
  Oceans Policy 36, 77
  opposition to marine protected areas 81–82
  Resource Management Act 1991 77–78
no-entry areas 63, 195
Nordic Forum on MPAs in Marine Spatial Planning 106, 107
North American Commission on Environmental Cooperation 11
North Pacific Regional Fisheries Management Council 17, 34

Norway 98, 99, 106–107, 168
no-take areas 28, 31, 48, 171, 195
  Great Barrier Reef Marine Park 60, 62, 63, 64, 65

Obama, Barack, administration 159, 161, 169, 178
Ocean Policy Task Force (US) 159–163
Oslo Convention 99
OSPAR (Oslo-Paris) regional agreement 8, 98–101, 108, 168
overexploitation 30
oxygen levels 4

Pacific Fisheries Council Scientific and Statistical Committee (SSC) 28–29
Pacific Marine Analysis and Research Association (PACMARA) 151
Paris Convention 99–100
participatory management 179–180, 193–195
participatory planning 15, 44, 46, 167
  Great Barrier Reef Marine Park 64, 65–66
  Massachusetts 156–157
  Namibia 143, 145
  RAMPAO initiative 137
  United Kingdom 93
Particularly Sensitive Sea Areas (PSSAs) 127, 128
pelagic protected areas 33
Pelagos Sanctuary for Mediterranean Marine Mammals 33, 54, 122, 130–131
place, sense of xii, xiii, 47, 54
planning ocean zoning 45–50
  Asinara Marine National Park 112–118
  see also implementing the plan
planning tools 46–47, 48, 49–50, 65, 151, 161
  when and how to use 169, 171–172
PNCIMA 150–151
pollution 3, 47, 185–186
population pressure on coastal zones 4–5, 131
priority area identification tools 46–47, 48, 49–50, 65
private sector engagement 46, 56, 141–142, 179, 180–182, 195

process of ocean zoning 15–16, 42–43, 173, 193
  MPAs as a starting point 51–56, 89, 129
  planning 45–50, 112–118
  visioning and goal-setting 43–44
  see also implementing the plan
productivity of fisheries 27–28, 29
property rights 9–10, 26
  commons 34, 35, 44, 181, 183, 194
  indigenous peoples 81
  New Zealand fisheries 78
PSSAs (Particularly Sensitive Sea Areas) 127, 128
public consultation see participatory planning
public misconceptions and resistance 34–35
  Great Barrier Reef Marine Park 68–69

Queensland 61, 66, 69, 167

RAMPAO initiative 136–138
RAMSAR Convention on the protection of wetlands 25
recognition of need 43
regional agreements 178
  Helsinki Commission 102
  OSPAR 8, 98–101, 108, 168
  shared resources 19, 26, 187–188
regional planning initiatives 7
  Irish Sea Pilot 84–88
  RAMPAO 136–138
regional zoning approaches 10–11, 159
research sites 52–53, 195
resource allocation for planning process 43
Resource Management Act 1991 (NZ) 77–78
restoration of key areas 186
Rhode Island 8
rights 34, 35, 192
  see also property rights
River Basin Management Plans (UK) 94
Rufiji Delta 139

sacrificial areas 18, 51, 69
St Lucia 28
Sanchirico, J.N. 26, 32–33
Sardinia 112, 113

scale of management interventions 24–26
Scottish National Heritage 91, 92
Seaflower Marine Protected Area 70, 71–72
shipping lanes 107, 124
single species management 27, 185
situation analysis 43
social-economic-cultural operating principles, Great Barrier Reef Marine Park 64, 65
South Australia 71
Southwest Indian Ocean Fisheries Project (SWIOFP) 142
spatial analysis 46–47
Spatial Multiple Criteria Analysis (SMCA) 113, 114–116
Specially Protected Areas of Mediterranean Importance (SPAMI) 122, 124, 127, 129
Special Management Areas (SMAs) 67
Sperrgebiet National Park 143
spillover 27–28, 29, 30, 182
stakeholders 44, 46, 54, 193–194
 governance 179, 180
 see also participatory planning; user groups
stock effect 27–28
Stratton Commission 161
SWIOFP (Southwest Indian Ocean Fisheries Project) 142
Symmens, Owen 78, 79–80

Tanzania 139–142, 182
Tanzania Marine and Coastal Environmental Management Project 140–141
territorial user right fisheries (TURFs) 182
Tethys Research Institute 131
time horizons 35
tourism 31, 140
 British Columbia 148
 Chumbe Island Coral Park 141–142
 Great Barrier Reef Marine Park 61
 Mediterranean 131
trading zones 181, 192
Tribal Parks 153
Tunesi, Leonardo 113

umbrella species 55

United Kingdom (UK) 8, 84
 identification of sites for protected zones 91–92
 Irish Sea Conservation Zones project 88–89
 Irish Sea Pilot 84–88
 Marine and Coastal Access Act 2009 84, 86, 89–91, 94, 167
 Scottish marine planning 91, 92
United Nations (UN)
 Biosphere Reserve programme 18, 34, 71, 177–178
 Convention on Biological Diversity (CBD) 125
 Convention on the Law of the Sea 24–25
 Environment Programme (UNEP)
  Global Programme of Action 11
  Regional Seas Programme 124
 International Maritime Organization (IMO) 124, 127, 128
 Regional Seas Conventions and Action Plans 24–25
United States (US) 8–10, 36, 156
 Everglades restoration 186
 Federal ocean policy 34–35, 159–163, 169, 178
 Massachusetts Oceans Act 14–15, 156–158
 National Marine Fisheries Service (NMFS) 33, 53
user groups 7, 26, 194
 education 183–184
 fisheries *see* displacement of fisheries
 misconceptions and apprehensions 34–35, 68–69
 visioning 44
 see also civil society governance

Vancouver Island 151, 153–154
van Vugt, Mark 44
Vietnam 8
Villa, F. et al. 114, 116–117
visioning 44
volunteers 180

Wallace, C. and Weber, B. 78
Washington (state) 180
watershed and coastal zone management linkage 12, 24, 25–26, 167, 184, 187

United Kingdom 94
Vancouver island 153
wealth inequalities 4
West Africa
    Bijagos Biosphere Reserve 135–136
    RAMPAO initiative 134, 136–138
West Coast Aquatics 153
Western English Channel fisheries exclusion zone 28
Western Indian Ocean Marine Science Association (WIOMSA) 142
wetlands 185
Whakatohea people 81
wind farms 104–105, 157
work plans 43
World Wildlife Fund (WWF) 151

Zanzibar 141–142
zoning 6, 7–8, 17, 166
    adapting from land to oceans 9–10, 26, 34, 102
    as an integrator 11–13, 14, 19, 25–26
    countering public misconceptions 33–35, 68–69
    in dynamic environments 32–33, 129
    European Commission principles 12–13, 169, 170–171
    lessons derived from case-studies 166–169
    Mafia Island Marine Park (MIMP) 140
    Namibian Islands MPA 143
    in Norway 106–107
    potential in the Mediterranean 124, 129
    principles 172–173
    tools *see* planning tools
    see also Comprehensive Ocean Zoning (COZ); Great Barrier Reef Marine Park (GBRMP)